WHAT PRICE COD?

A Tugmaster's View of The Cod Wars

by Norman Storey

HUTTON PRESS
1992

Published by

The Hutton Press Ltd.,
130 Canada Drive, Cherry Burton, Beverley,
North Humberside, HU17 7SB

Printed and bound by

Clifford Ward & Co. (Bridlington) Ltd.
55 West Street, Bridlington, East Yorkshire,
YO15 3DZ

ISBN 1 872167 44 6

CONTENTS

ACKNOWLEDGEMENTS

Sincere thanks to Elaine (Honey Bun) for her ability and patience deciphering my hieroglyphics.

Thanks to Mr. Dag Pike for the pick and the use of his photographs.

Thanks to Mr. Albert Smith for the use of his action photographs.

Thanks to Mr. Archie Macphee for his historiography and his story in *The Listener* December 1975.

Thanks to Skipper Dickie Taylor for his hydrographic talent on the reproduced chart.

APOLOGIES

For any discrepancy in dates, names and places, the following was written from memory after a lapse of 18 years and more.

INTRODUCTION

The Cod War between Iceland and Britain has now passed into history, some would say ignominiously, for it is a dispute that should never have come about. It ended in defeat for Britain after years of bitter wrangling.

Iceland wanted the sole rights to all fish stocks around her shores, first in 1952 up to 4 miles and finally in 1976 up to 200 miles. Britain protested strongly, with the backing of the International Court of Justice in the Hague.

Britain's case was that the sea outside territorial limits (3 miles from base lines in 1952) was free for all to navigate and to use for fishing. Iceland, ignoring the finer points of international law, maintained that she had the sovereign right to establish fishing limits where she pleased. With each new regulation extending fishing limits, Iceland adopted a policy of harassing British trawlers fishing inside the unilaterally declared limits.

Icelandic coastguard vessels at first attempted to arrest "offending" trawlers, but when this failed, they cut trawl wires without warning thereby creating heavy losses for the fishermen. The trawling industry asked the Government for protection and the Royal Navy was called in to prevent the harassment of the trawlers. So the diplomatic dispute became a sea "war" with all its inherent dangers.

The good sense and skills of the British sea commanders meant there was no shooting and few casualties were recorded other than to pride and loss of fishing gear. It was not until the later stages that tempers frayed and an Icelandic gunboat fired a shot at one of the British protection vessels and missed.

Before they begin this story of the Cod Wars, readers are recommended to glance at the following historiography. It will act as a guide and give them a better understanding of a complicated and devious affair.

HISTORIOGRAPHY

1822: North Sea Fisheries Convention, signed in the Hague by France, Germany, the Netherlands, Denmark and Britain, establishes a 3-mile fishing limit in the various regions.

1901: British-Danish Convention signed establishing a 3-mile fisheries limit around Iceland, the Faroes and Greenland.

1944: Founding of the Republic of Iceland which becomes a free state and no longer a colonial dependency of Denmark.

1949: Iceland serves two years' notice of her abrogation of the 1901 British-Danish Convention on fishing limits around Iceland.

1952: Iceland extends her fishing limits from 3 to 4 miles; Britain protests; British fishing interests impose landing ban on imported Icelandic fish.

1953: Iceland signs long-term agreement with Soviet Union exchanging fish and fish products for 90 per cent of her needs of oil.

1958: Iceland extends her fishing limits from 4 to 12 miles; Britain declares act illegal under international law and sends in Royal Navy frigates to prevent enforcement of the new regulations by Icelandic Coastguard vessels.

1960: Britain withdraws frigates pending outcome in Geneva of Second UN Conference on Law of the Sea and negotiations on settlement of Anglo-Icelandic dispute.

1961: Iceland and Britain reach agreement; British trawlers allowed to fish inside 12-mile limit for trial period of 3 years when any further extension to be taken to the International Court of Justice; normal relations and trade between the two countries resumed.

1964: Belgium, Denmark, France, West Germany, Ireland, Italy, the Netherlands, Portugal, Spain,

Sweden and United Kingdom agree in the European Fisheries Convention to establish a 12-mile fishing limit with a concession for each of them to fish up to 6 miles off each other's shores on the basis of a ten-year historical right.

1971: Announcement by new left-wing Icelandic Government that it intends to abrogate the 1961 Anglo-Icelandic agreement and to introduce an extension of the limit to 50 miles.

1972: Iceland introduces legislation increasing limits from 12 to 50 miles; Britain protests and refers matter to International Court of Justice; Iceland refuses to accept jurisdiction of the Court in favour of Britain and disagrees with its findings.

1973: Iceland introduces measures to enforce new limit; British trawlers warned to withdraw to 50-mile line or have their trawl wires cut; 60 British trawlers lose their trawls in incidents with Icelandic coastguard vessels; four British ocean-going tugs sent as protection vessels followed by Royal Navy frigates; later defence tugs and frigates withdraw pending outcome of peace talks in London.

1974: Iceland and Britain settle their differences; Britain accepts case for 50-mile limit in return for 139 registered British trawlers being allowed to fish in restricted waters up to a total annual catch of 130,000 tons and the agreement to last two years.

1975: Iceland increases limits to 200 miles; defence tugs and frigates return to Icelandic waters to protect British trawlers against trawl-cutting activities of Iceland's Coastguard; later Iceland complains to United Nations Security Council on use of force by United Kingdom; Britain replies that the role of her vessels was solely to defend British trawlers against attacks by Icelandic Coastguard.

1976: Iceland breaks off diplomatic relations with Britain on 19 February: NATO calls meeting in Oslo between foreign ministers of the two countries on 31 May: It was agreed that British trawlers would be allowed to fish in Icelandic waters to a catch of 50,000 tons; that only 24 trawlers from a registered list of 93 would be allowed to fish at any one time; and that the Agreement was for six months only.

1976: Anglo-Icelandic agreement reached in Oslo expires on 1 December; it is not renewed.

1976: Britain enacts her own 200-mile limit under the Fishery Limits Bill on 10 December.

AUTHOR'S COMMENTS

This chronicle of events, is yet another aspect of the last two "Cod Wars" as seen by myself, the Master of the support tug *Lloydsman.*

No criticism is intended. The following catalogue of events was witnessed by me and are mostly humorous, written for the benefit of seamen in general, specifically Deep Sea Trawlermen, who served in Icelandic waters at that time.

Captain Norman Storey OBE, ex marine/towing superintendent.

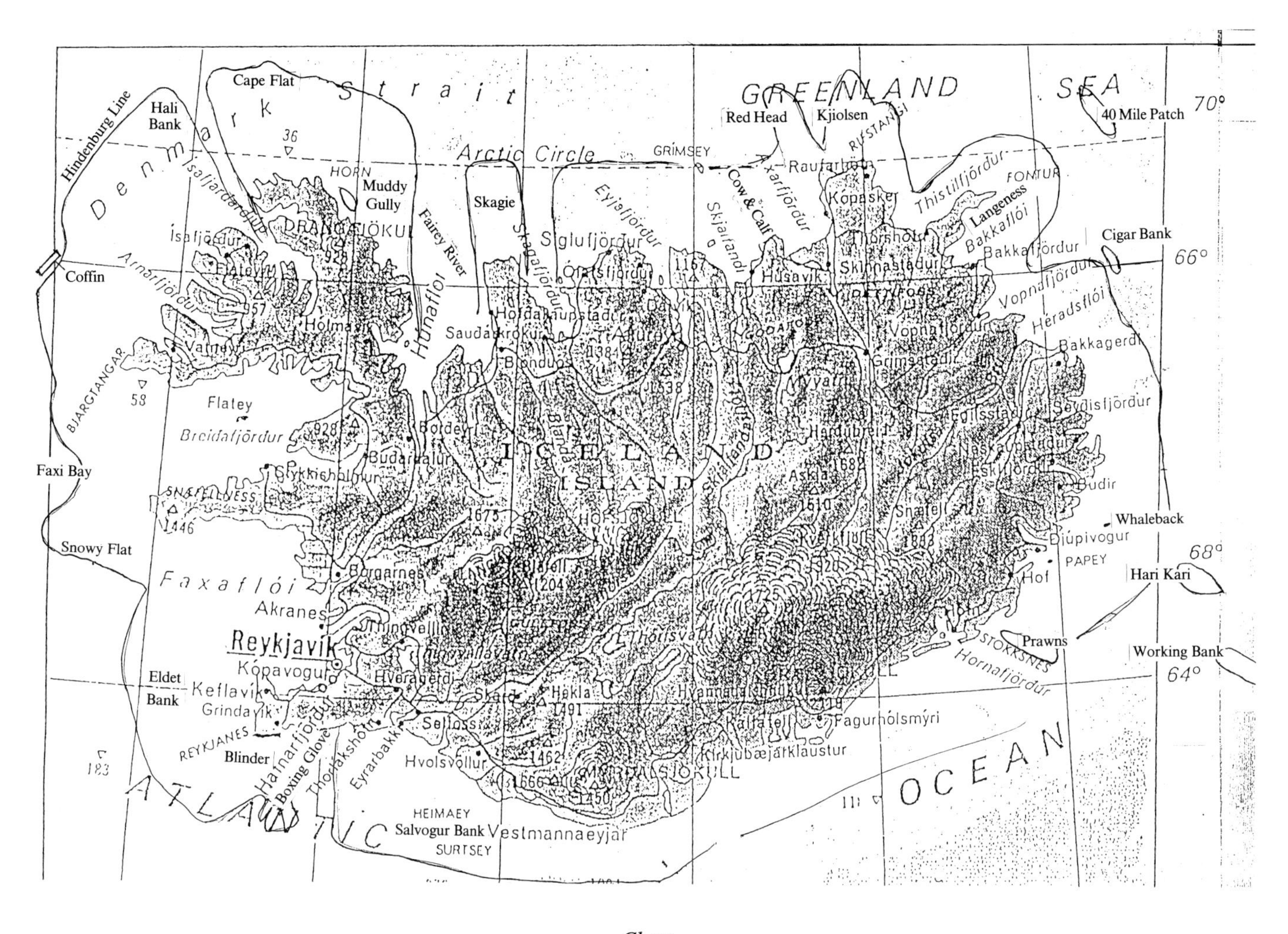
GREENLAND SEA
Denmark Strait
Arctic Circle
Cape Flat
Hali Bank
Hindenburg Line
Red Head
Kjiolsen
40 Mile Patch
70°
Muddy Gully
Skagie
Fairey River
Cow & Calf
Langeness
Cigar Bank
66°
Coffin
Faxi Bay
Snowy Flat
Whaleback
68°
Hari Kari
Prawns
Working Bank
64°
Eldet Bank
Blinder
Boxing Glove
Salvogur Bank
Reykjavik
Akranes
Faxaflói
Flatey
Breidafjördur
Húnaflói
Sauðárkrókur
Siglufjördur
Eyjafjördur
Húsavik
Raufarhöfn
Kópasker
Thistilfjördur
Bakkaflói
Bakkafjördur
Vopnafjördur
Héradsflói
Bakkagerdi
Seydisfjördur
ICELAND
ISLAND
Keflavik
Grindavik
Kópavogur
Hveragerdi
Selfoss
Hvolsvöllur
Hekla
Borgarnes
Kirkjubæjarklaustur
Fagurhólsmýri
Djúpivogur
PAPEY
Hof
STOKKSNES
Hornafjördur
REYKJANES
HEIMAEY
SURTSEY
Vestmannaeyjar
ATLANTIC
OCEAN
GRÍMSEY
HORN
Ísafjördur
Arnarfjördur
Vatneyri
Hólmavik
BJARGTANGAR
SNÆFELLNESS
Stykkishólmur
Thorlákshöfn
Eyrarbakki
Hafnarfjördur

Chart

CHAPTER 1

First Cod War 19.1.73 to 16.11.73

Tug *ENGLISHMAN*

Owners:	United Towing Limited
Built:	1965 Cochranes, Selby
Length:	136′ x Beam 33′ x Depth 13′ 6″
Gross:	574 Tonnes
Berths:	21 Men
B.H.P.:	3000
Speed:	14.5 Knots
Bollard Pull:	42 Tonnes
Master:	Capt. N. I. Storey

When the second disagreement over fishing rights started over the 12 mile limit, the first being the 3 mile limit, tugs did not participate then. The tug *Englishman* was sheltering in Stavanger Norway, in the middle of a tow of two ferries from Karleserona, Sweden to Glasgow.

We were released from this tow to the Ministry of Agriculture Food and Fisheries, and ordered to Hull for stores and briefing.

Reason for the dispute: The second dispute concerned the 12 mile limit. The Icelanders claimed the British trawlers were over-fishing the breeding grounds, taking 150,000 tonnes per year which they considered was excessive.

The Labour Government in its wisdom with Mr. Callaghan as Foreign Minister and Mr. Hattersley as Minister of Agriculture Food and Fisheries, thought GUN-BOAT DIPLOMACY instead of negotiations was the way to play it. Hence the fiasco of the 1st, 2nd and 3rd Cod Wars.

Instructions: Apparently in the 3 mile dispute the Icelandic Coastguard Vessels (I.C.V's) would tow their cutters at the trawlers from astern, crossing the warps very close to the trawler, hooking the warps and using their weight to cut the warps.

My orders were to get between the trawler and the I.C.V. to stop him passing over the fishing gear. On no account must we collide, 'rules of the road' being strictly observed.

The *Englishman* had a speed of 12 knots and the I.C.V's were 18 knots. I could protect one maybe two trawlers but what about the other 150 or so trawlers fishing Iceland from Hull, Grimsby, Fleetwood and Aberdeen?

Each tug would have a representative from M.A.F.F. on board, called a Defence Commander, to ensure we played the game to M.A.F.F. rules.

Late January 1973: Approaching the Whaleback Fishing Grounds (SE corner Iceland see Chart) I made contact with *Ranger Brisies*, an ex-trawler, now a support ship for the U.K. fishing fleet. Two trawlers were laid unable to fish because the I.C.V. *Odinn* was laid with them. That's all he had to do, lay and fishing stopped.

It was a beautiful afternoon, no wind, sunny and a glass calm sea. I decided to throw the gauntlet down and see what happened. I steamed the *Englishman* close across the I.C.V. *Odinn*'s bow, close enough to make him go astern, circled him close with my crew on deck, jeering.

Then full speed to the trawlers, I asked them to shoot their gear, one was willing and was just "blocking up" when I got there. As soon as he started towing I was between his warps, my stem some 20 feet from his stern.

Immediately, the I.C.V. *Odinn* started towards us. When he passed the *Ranger Brisies*, Paddy Donavan, the Master, told me he had his cutter down.

There was I, in the place I was told to be, over the warps close to the trawler; that was the place they cut the warps, and I must not be conned out of position allowing him in close to the trawler's stern. I was not happy.

I had discussed this a hundred times with my first and second mates, while running off to the grounds in the 3

The 'Englishman'

and a half days it took. If I was in the gun boat and the tug stayed in this position, no matter what he did I would be certain I could get the net or warps. Why do I have to use a short cutter? Steam up close to the tug's stern with a long cutter. When down to 6 knots towing speed the cutter must be close to the net, steaming at 45° across the stern of the tug and up the side of the trawler would get the warps; also dropping the cutter on the bottom, on to the bow of the trawler, at the last minute moving on the other bow, leaving the cutter on the other, when the trawler passed my stern, steam down the side and across his stern (looping).

All these theories would work if the tug would stay put.

I decided I was not going to stay in one position. The I.C.V. *Odinn* was heading straight for my stern at 18 knots. I let the *Englishman* drop astern of the trawler about 100 feet, then the I.C.V. *Odinn* shaped to pass me on my starboard side. When I was sure he would come starboard I increased to full (12 knots) and kept between him and the trawler. By the time I was abeam of the trawler, the I.C.V. was half a length past me but I held my line forcing him well ahead of me and the trawler. I am an ocean tug with plenty of tyre fenders. He thinks, I hope, that I might ram him. When I was ahead of the trawler, I eased back staying only about 50 feet ahead of the trawler, always careful not to get conned out of position.

Sure enough when he was about half a mile ahead he turned hard to port and came down the port side of the trawler at 18 knots plus 6 knots the trawler's speed = 24 knots over the ground towards each other. Decision time again, am I out of position?

I moved fine on the trawler's port bow, stopped, letting the trawler pass me and manoeuvred keeping my bow straight at the I.C.V. By the time he passed the cutting point I was there, although wrong way around heading at him, stern to my charge. He was well clear and I could see his cutter on the surface because of his speed. Again I thought — don't get conned, so back to the stern of the trawler.

It would appear that providing I stayed between the I.C.V. and my charge, we had him covered. The I.C.V. *Odinn* then proceeded about 5 miles off, then laid. The other trawler had decided to shoot after watching all this. I asked him to stay close, but trawler skippers have a mind of their own and he towed about 3 miles away.

When the two trawlers and the I.C.V. were in a triangle, all of 5 miles apart, in he came again, full speed. I was halfway between the two trawlers, unsure which one to go to. This was when I first appreciated their skill and seamanship. He knew that the longer I hesitated the closer he got to me, which gave him a better chance to outrun me to either or both trawlers. Should I sacrifice one to go to the other or what? I called on the very high frequency (VHF) radio and told them both to haul, which they did without question. The I.C.V. steamed up to me and laid.

Back to square one.

I went back to the trawler which first shot his gear. He was still willing to "have a go" so he shot his gear and I escorted him. The other just laid with the I.C.V. *Odinn*. At dawn the I.C.V. *Odinn* left, sailing over the 12 mile line.

Bad weather was coming. It blew for a week from the west and fishing was sparse. The I.C.V.s never bothered us when the fishing was poor. They obviously monitored the radios. The trawler skippers were never off the VHF making if easy for the I.C.V's to locate us.

We heard on the world news that an Icelandic trawler was missing, presumed sunk. The I.C.V. *Aegir* was out searching, and two naval frigates, 2 trawler support ships and two tugs all volunteered to assist. All in line searching night and day for a week under the orders of the *Aegir* and the frigates.

21 February 1973: The *Englishman* developed rudder seal trouble and we went to Skaula Fjord, Faroes for repairs. We heard the seaching fleet had found one badly weather beaten life raft with one body tied by the wrist to the raft.

Thorshaven was a bit anti to us but Skaula, an island where the shipyard was, was friendly.

The I.C.V. 'Odinn'.

The Faroes, being Danish, lent towards Iceland in the Fishing Dispute.

We were up a slipway with an 18 foot ladder for a gangway. The foul weather continued with wind and snow. The crew would not venture ashore but the local children had no problem climbing up the ladder to visit. Although none could speak English their little smiling faces said it all and they were made very welcome.

There was a weekly Mobile film show, ashore. The tug had 3 films on board, an Italian Robin Hood, a 007 film and "Mountain Man". All were shown every night on the tug to the local children and later their parents.

In less than 1 week we were out of coke, chocolate and beer but very friendly with the natives. Except the local film mogul.

2nd March 1973: We lost both engines because of air locks when we launched off the slip, fouled a mooring buoy but cleared it with our own divers. The repairs had taken about 13 days to complete, slow but thorough.

5th March 1973: When I returned to the Whaleback everyone was conspicuous by their absence, the VHF silent on all channels. I had an ex-trawler R/O on board who went to work on the "big set" and in a short time handed me a couple of grid chart references. These applied to the grid charts which the Ministry of Agriculture, Food and Fisheries had issued everyone with, supposedly so the navy would know where everyone was (the skippers had some weird and wonderful names for all the fishing grounds — see Chart) and hoping to confuse the I.C.V's. We found out later all the I.C.V's had copies of these grid charts.

All the fleet had gone to the Northside or the NE corner. 36 hours later we found most of the fleet, about 60 miles round the NE corner on the North side at "Red Head'.

As we arrived so did the I.C.V's, two of them. They had left the trawlers alone for two weeks after we (the support ships) had assisted in the search for the Icelandic trawler that was lost, plus another 10 days while talks were going on between the two governments, but they failed, so things were back to normal.

CHAPTER 2

Tug *STATESMAN ex. ALICE MORAN*

Owners:	United Towing Limited.
Built:	1966 Kure Japan.
Length:	180′ x Beam 41′ x Depth 19′.
Gross:	1166.7 Tonnes.
Berths:	21 Men.
B.H.P.:	9600.
Speed:	18.0 Knots.
Bollard Pull:	100 Tonnes.
Master:	Charlie H. Noble.
Flag:	Panamanian.

Labour M.P's in the government objected strongly to a Panamanian Flag Ship defending British Trawlers. The tug was renamed *Statesman One* and her flag changed to the Red Duster.

The *Statesman* was already with the fleet, about 50 in number, spread over some 30 miles. To defend in the normal way was useless in this situation. The trawlers themselves gave us the answer. They were prepared to fish if the I.C.V's were far enough away. If the I.C.V's could be kept on the move then the trawlers would look after each other, fishing in twos and threes, with one on guard. So *Statesman* and *Englishman* spent the next 35 hours chasing, trying to head off and setting collision courses.

I remember asking a certain skipper if I could pass across his head to save time and distance. He had other trawlers towing parallel to his course. He said yes and gave me a lot of Yorkshire verbal encouragement. When I was past and clear I heard his voice again, "Who is the silly ... that's just run into me?" They were both so interested in watching me they had closed together and bumped.

When the I.C.V's finally left, the fleet was down to about 12, all in a tight bunch. Most were going home after this haul. Two sets of gear had been lost. I went close to *Statesman* to discuss the situation on the loud hailer, not on the radio for obvious reasons. The I.C.V's had gone east so maybe the boys had gone west to the NW Cape. He suggested because I was the slower I should go and he would follow when this bunch of trawlers went home. I agreed and "West" we went.

We found them at Fairy River, about 35 miles east of the NW corner. Within a couple of hours the glass had dropped, the wind was from the north and the black frost fog was rolling in.

I was the stranger here so I followed the pack. Everybody, some quicker than others, hauled their gear and made the best speed they could round the corner up the fjord, under Ritta Head. I was there in 4 hours yet the whole of my after deck was full of solid ice, the winches and bollards solid in enormous blocks of ice. 1″ wire rigging was now 7″ across, an eerie grotesque fearful sight. We spent two days chipping ice from the rigging to try to get the weight down.

The fjord was too deep to anchor, not that anyone would dare for fear of being arrested. Everyone just steamed easy to the lee of the Cliffs then drifted off, then back again, constantly watching the port end of the fjord for I.C.V's.

I laid alongside the *Kingston Pearl* skippered by a Withernsea friend of mine Jack Trip, who had offered me a fry of fish and a yarn.

That old boy knew the elements around that part of the world. He said the glass was going up slowly, the temperature would soon and we'd be away in 4 hours, we were too.

When I let go of him to drift, my bulwarks hooked under his towing block, a patented gadget they put the warps in when towing, to keep them away from the propeller. The gentle swell had lifted the block out of his bulwarks and it stayed balanced on my bulwarks. When I took it back the healing leg was broken. Within minutes Paddy, skipper of the *Ranger Brisies*, had his boat down took the block on board his ship, welded it and had it back before sailing time. Just one of the many duties carried out by the three support ships for the British fleet at Iceland. *Ranger Brisies*. *Othello*, both converted trawlers, and the converted Norwegian Schooner *Miranda*.

Because it was not allowed into Icelandic ports, the fleet had to be as near self-contained as possible. Each support ship had a doctor and hospital on board, first class engineers, mechanical and electrical, (only twice in the two Cod wars had a trawler to be towed home) plus some of the finest small boat crews I have ever had the pleasure of working with. No weather was too bad and no errand too trivial.

As suspected, when the weather moderated, the I.C.V. *Aegir* appeared out of the little port. He did not charge about the fjord or even call on the VHF, he just appeared as if to say the weather is okay now out you go, and out we went.

He went south, we all went north and east. Each trawler seemed to go to a different ground, whatever took their fancy. I stayed with the most westerly group thinking that's where the danger lay, behind us to the west.

We gradually worked our way east. Nobody found anywhere to warrant a second tow. We finished back at the NE corner Red Head Bank where about 30 boats had congregated, *Statesman* with them. We had noticed the Icelandic Coastguard plane overhead each day since leaving the NW cape.

All the trawlers were towing up and down the same 3 mile tow, almost line astern. It was a beautiful day with no sign of the I.C.V's.

My Chief informed me he had a job on the starboard engine that would take about 4 hours on "one leg". I decided to steam at half speed on my port engine slowly round the pack and keep quiet about being on one leg. Fortunately I chose to go west first. As soon as I arrived at the end of the pack I could see an unknown vessel about 3 miles away coming towards the pack I did not know the ship, but the trawlers did and I was soon informed, it was the I.C.V. *Albert* an old ship of about 12 knots.

Motor Tug 'Statesman' 12,000 h.p.

The 'Statesman', after being re-named 'Statesman I'.

My Chief told me it would take as long to put the starboard engine back together as it would to carry on with the repair. I told him to carry on.

The I.C.V. *Albert* stopped and laid about a mile away from the fleet, I knew *Statesman* was 5 miles away at the other end of the fleet, only 20 minutes away. I edged over to him and laid close by, reporting to the fleet that the I.C.V. *Albert's* cutter was up and I would report his movements. The VHF was full of advice, trawlermen being what they are, i.e. if I could not stop him they would. They all had more speed than him. He was certainly listening to all of this. I thought perhaps he was just observing and reporting the catches. What we did not know was the I.C.V. *Thor* was 30 miles east coming at full speed.

After about 45 minutes he got under way. At about 9 knots on one leg I had no trouble keeping up with him close to his stern. I must state he had nerve. He held his course through and along the fleet to a lot more verbal abuse on the VHF. Half-way through I got my starboard engine back. When he cleared the pack he altered course for the 12 mile limit and home. The I.C.V. *Thor* arrived and started his antics. With two tugs chasing him he seemed to lose interest and soon left to the west.

I was due replenishment in a few days so I went south to the Whaleback for my last few days. We knew the I.C.V.*Odinn* was bothering the boys there.

When I arrived there were no I.C.V's, but the boys had found some fish about 100 to 150 baskets a haul. Within two days the whole fleet was back about 60 trawlers and of course out came I.C.V. *Odinn* and I.C.V. *Thor*.

Their tactics were slightly different this time. They seemed to make more concentrated efforts towards the most isolated trawlers. *Statesman* and I decided to revert back to trying to look after one or two each. The pack had scattered over 5 miles, but the effect had reversed. The I.C.V's concentrated on the ones we were with and left the fleet alone. The I.C.V. *Thor* decided he wanted the trawler I was guarding. For two hours he ran around us coming in from every angle. I had to really threaten to ram him to try to deter him. Then something new happened. Someone appeared on the wing of his bridge and levelled a rifle straight at me. I shouted to the mate, look at that. He said he had been there 10 minutes. I had been too busy and I had not noticed. Purely on the theory if he had not fired yet, he wasn't going to, I ignored him.

Unbeknown to me I.C.V. *Odinn* had left *Statesman*, so Peter came to give me a hand. I.C.V. *Thor* and I.C.V. *Odinn* left the area, and my trawler hauled 150 baskets, all well worthwhile.

The trawler I had been guarding told me he was going for prawns, at Stokksnes, (see Chart), his last haul before he went home. There would be a fry for *Statesman* and I , if we stayed with him. As I was due to go home also, I agreed, *Statesman* said he would stay as long as he could. The weather had deteriorated when the trawler hauled and the skipper said he would buoy two bags of prawns if we could pick them up. No problem!

I went in first for the two bags, one for *Statesman*. I had the whole crew ready on the deck. Both sacks were emptied, all the prawns headed and tailed. We filled *one* sack with the heads and tails, added a box of kippers and rebuoyed the sack. I told *Statesman* his share was on the buoy and waited for the explosion on the VHF. When it came, Pete the Master excelled. It took 30 minutes for him to give me my character reference without repeating a single swear word twice.

The trawler skipper was totally mystified by this so I had to explain the trick I had played on Pete. He saw the funny side to it and promptly buoyed two more sacks for Pete.

I left the grounds then for Lerwick, Shetlands, for replenishment and my relief. A nine week voyage for the first trip.

Arrived 27th March 1973.

CHAPTER 3

27.4.73 to 26.5.73

Tug *IRISHMAN*

Owners:	United Towing Limited.
Built:	1967 Cochranes, Selby.
Length:	128′ x Beam 32′ x Depth 14′.
Gross:	452 Tonnes.
Berths:	16 Men.
B.H.P.:	3000.
Speed:	13 Knots.
Bollard Pull:	30 Tonnes.
Master:	S. Hawkins.

The Ministry of Agriculture, Food and Fisheries requested another tug for Iceland to assist *Englishman* and *Statesman*. The tug *Irishman* was chartered. I joined her in Hull as Second Master to complete half the voyage and assist the master with Navy and M.A.F.F. procedure because I had some experience there, and he had not been briefed.

We sailed from Hull to the Whaleback SE Iceland. On arrival things were very quiet. The two governments were talking and while diplomatic negotiations were going on the I.C.V's stayed inside the 12 mile limit lines. The only incident in this period was when an Icelandic trawler persisted in fishing very close to our fleet. The comments on the VHF were plentiful e.g.

"Chase him off Norman."

"I am not fish finding for an Icelandic web-foot."

"When I haul this time I'll ram the ..."

I was ordered to stay between him and our boys, ironically the opposite to my charter. Instead of defending our boys from I.C.V's, now I was defending an Icelandic trawler from our boys.

For several hours I carried out my orders trying to contact him by VHF or loud hailer when I was very close to him. All to no avail. After about eight hours he hauled and left and an English voice said thank you and wished me all the best. The Icelander had been chartered by I.T.V. and were just recording the VHF channels for footage for the I.T.V. news programme, but as he told me by the time they had bleeped out the bad language, there was nothing left.

I left the *Irishman*, after a five week voyage, in Lerwick, Shetlands, where he replenished and returned to the fishing grounds.

CHAPTER 4

7th June 1973 I rejoined my favourite tug *Lloydsman.*

Tug *LLOYDSMAN*

Owners:	United Towing Limited.
Built:	1971 at ROB Caledon Yard, Leith.
Length:	260′ x Beam 47′ x Draft 26′.
Gross:	2040 Tonnes.
Berths:	34 Men.
B.H.P.:	10,000.
Speed:	18 Knots.
Bollard Pull:	135 Tonnes.
Master:	Capt. N. I. Storey.

We sailed from Hull with orders to proceed to Rosyth Navy Yard and meet Admiral Lucy who wanted to brief me. I was anxious to say the least. After all you don't meet an Admiral every day.

I berthed the *Lloydsman* in the forenoon at the dockyard. A small tug had been sent to assist me berthing. I thought this amusing but did not use him. There must have been 30 men in the shore mooring party and when I enquired why so many? I was told the Big "L" was Britain's biggest tug and they were not sure what to expect. My chest went out another 6 inches that day. The

The 'Irishman'. Courtesy Walter Fussey & Son.

The 'Lloydsman'.

biggest gangway I had ever seen landed on the forecastle head. I swear we laid over a degree or two towards the quay with the weight.

The Admiral's aide, a Commander, came on board and told me the Admiral and his party would be on board at seven bells, 11.30 hours and left.

The Admiral and his entourage duly arrived all ten of them. While escorting him from the gangway to my day room he asked me if I would allow half his escort to look around my tug because he was short of time and they could report to him afterwards. Then he could brief me at the same time. I of course agreed and told my First and Second Mates to assist them.

Once in my day room with the introductions over, the Admiral obviously relaxed, accepted my liquid hospitality and we got down to the briefing that turned out to be the reverse, a debriefing from me to him, on my two previous visits to the Icelandic fishing grounds. They were very interested in how the I.C.V's attacked the warps, what angle they cut on, how they did a loop cut, how they laid with the cutter on the bottom and let the trawler tow over it. Also what the tugs and frigates could do to ward off these attacks bearing in mind the no-collision rule, what they could do if the no-collision rule was lifted, would the water cannon be effective, how best to use it. The discussion lasted about an hour. I noticed the aides taking notes.

When the meeting was over, the Admiral informed me the debriefing had been most helpful to his plans. Of course he did not tell me what they were.

He ordered me to sail at midnight, after the defence commander and a three-man B.B.C. TV team had boarded, He then invited me to visit the N.A.T.O. Defence Bunker close by, which I accepted and visited that afternoon. What I saw gave me a very reassuring feeling, because the bunker was covered by the Official secrets Act and that's all I can say about it.

My passengers boarded in good time. The Defence Commander was Bill Bridges an ex-Lowestoft trawler skipper and a good hand. Dave Whickam was the B.B.C. reporter with a camera man and sound engineer.

We duly sailed and had a good passage to the North West Iceland fishing grounds. The trawlers were well spread between the Coffin and Cape Flat (see Chart). We had not seen any I.C.V's all the voyage, the word from the fleet was the governments were talking again.

A lot of time was spent running with trawlers and the other ships for the benefit of the T.V. crew, trying to get footage for the B.B.C. News, and documentaries.

One foul day I heard the skipper of the Grimsby trawler *Real Madrid* calling the *Miranda*, Capt. Willis Brown, on the VHF. He needed the doctor on board. One of his crew had been badly injured on the winch. The doctor came on the air and after listening to the skipper explain the injuries, agreed the man could not get in the sea-rider (a high powered rubber boat), so he would go to the man.

The two ships confirmed their position and started to steam towards each other.

I plotted both positions to find I was roughly between them. I called the T.V. crew and told them if they could use their gear in this awful weather they would catch a boat being launched on an errand of mercy, in a full gale and 20 foot sea and swell. It was amazing how fast those men could move. In no time at all, the bridge was cluttered up with gear and cable, the camera in a shoulder crutch, the whole lot covered in water proofs. The poor fellow could hardly stand up, what with the wind and the motion of the *Lloydsman*.

They were about 20 minutes early for the rendezvous so we dragged the camera-man back to the bridge again. When I saw the trawler coming I called the skipper and told him that we were going to put the whole manoeuvre on the B.B.C. News and could he pass me close for the camera. He obliged and gave the camera crew some excellent, exciting footage. He went past me at the best speed he dare in that weather. All we could see for half the time were his masts and bridge. The rest was lost in the spray of the broken water.

I followed close and watched the boat launched off

the lee side of the *Miranda*, followed it across to the trawler and watched it trying to get alongside the *Real Madrid*. In those heavy seas, the skipper tried to lay across the sea to give the boat a lee but she was rolling far too much. He tried it head to sea but the sea rider was lifting 20 feet. I thought it was going aboard the trawler's "well" deck. At 45 degrees off the wind it looked best and the sea rider went alongside, still a very big lift, so much so that when the boat was up on the crest of the wave the doctor just stepped across to the trawler's bulwarks, an excellent job. 30 minutes later the doctor "jumped" back on the boat and went back to his ship, all in a day's work. The injured man was transferred to the *Miranda*'s hospital two days later, when the weather moderated.

The camera crew were delighted with the shots they had and hardly needed me to tell them that in my opinion they had witnessed as fine a piece of seamanship they would ever see. David Whickam the reporter did the sound track to accompany the film and I am pleased to say he stated that fact.

Shortly after this incident the *Statesman* appeared on the scene. Dave Whickam B.B.C. asked if *Statesman* could steam by the *Lloydsman* like the *Real Madrid* had, I asked Charles Henry the Master, who readily agreed when I told him the tug would be on the 6 o'clock news someday.

The weather was slowly moderating but there should be plenty of spray tossed upwards. I put the Big "L" on slow speed, head to wind and weather.

The photographer was ready. *Statesman* was coming towards me from aft to foreward doing the best speed he could in the "scruffy" weather, plenty of spray going over him, perfect. I watched him hold the *Statesman* in shot until her bow was a beam of my bridge, then he panned to the boat deck, then to the aft towing deck and on to the tug's wake. The camera man's comments ruined the sound track. I was watching when he panned from deck to wake and back quickly. Four of her crew were doing a "moonie" for us on the engine room hatch. They ruined the film, but I shall always remember the amusement it caused.

When the storm died down, things were back to normal, watching the trawlers fishing, cruising, trying to find something new to film.

A trawler reported he had fouled his propeller with a rope, and was at the Muddy Gully. I steamed over to him. My 2nd Officer and an A/B were qualified shallow water skin divers, part of a salvage tug's crew. I sent them over in our 6 man D.O.T.I. boat with a hacksaw each to see if they could do anything. Ten minutes sawing by both men and it was clear.

The skipper showed his appreciation by filling the boat with jumbo haddocks, so much so that one of my divers had to swim back, he could not get in the boat. Much appreciated.

The B.B.C. team had been onboard 2 weeks and were getting bored. The talks were still on. They asked if I could help to get them and their film back to U.K. I went on the VHF and asked if anybody was going home in the near future. A Hull trawler answered, he had two more hauls to make and he would take the three men and their gear.

I rendezvoused and ferried them over at 02.00 hrs on the 21st June 1973. At 08.00 hrs I collided with the I.C.V. *Odinn*. They missed it by 6 hours.

CHAPTER 5

The I.C.V. *Odinn* Incident

It was a beautiful morning, clear and sunny, with no wind and a glass calm sea. The watch called me to say the Navy was calling me on the portable ultra high frequency (UHF) set (on loan from the Navy). My First Mate said that he had heard on the world news, B.B.C., that the talks had broken down again, so I guessed what the call was about. It was the I.C.V. *Odinn*. He was about 5 miles SW of the trawler fleet, speed 12 knots and escorted by two frigates.

We had a good radar picture, the three dots were 5 miles from me and I was right on the line to the first trawler. My compliments to the watch, my night orders were to keep SW of the fleet. I ordered the second engine clutched in and all dead lights and water tight doors shut, this was normal practice but it might have caused my Defence Commander to ask the question: "What's your intention Norman?" I replied, "Let's see what happens, perhaps the Navy have fresh orders. Escorting the I.C.V's in is new." The mate informed me the I.C.V. *Odinn* was still coming straight at us. I ordered to keep him more than 2 points abaft the starboard beam, keep him the overtaking vessel. When he was two miles off he reduced to 8 knots but kept coming. At a half mile off me his bow wave dropped, apparently running his way off but still straight at me.

During the previous few minutes everybody on my bridge had commented as to what they thought the intentions of the I.C.V. *Odinn* were. Most of my colleagues thought he had come out to provoke the Navy or us into some incident or other. My money was on him acting as a decoy, to come out and lay with us, keeping three defence ships occupied while another I.C.V. attacked the other end of the fleet. I was just about to call the trawlers and tell them to keep a sharp look out to the east when the I.C.V. *Odinn* blew three blasts on his whistle. Whatever his intentions he was getting too close to me. I put the telegraph full speed ahead. If he was going to hit me, better aft (my strongest construction) not midships. The two ships collided, my starboard quarter and his stem. I stopped the *Lloydsman* as soon as we were clear and layed. I could not believe the damage the I.C.V. *Odinn* had received. We had felt only the slightest bump, (the watch below would not believe there had been a collision at all), yet six foot of the I.C.V. *Odinn*'s bow, about 2 foot above the water line had been rolled to starboard, like an opened sardine can, even though she had an ice reinforcement bow.

Thinking, if that's what has happened to him, what damage have I got, I sent the 1st Officer and the Chief Engineer to find out. Their report was, there was not a mark on the rubber belting that went all around the tug.

The bulwarks were set in 2 feet over 6 feet. Due to the rake of the I.C.V. *Odinn*'s bow as he over rode us, this had pushed one of the bulwark stanchions downwards splitting the top of the full fuel tank and there was fuel on deck.

How relieved I was. It could have been a hell of a lot worse.

A quick discussion with my Chief Engineer, A. Whiteley, and we decided to list the *Lloydsman* to port a little to stop any more gas oil running out on deck, transfer the fuel to another tank, fill the damaged tank with water and forget it until the next replenishment port.

The Defence Commander told us we were both ordered to report on a frigate, its motor boat was on its way to pick us up. In the rush to get ready I forgot the log book and had to come back for it.

The First Lieutenant met us at the gangway and took us to the operations room down below decks. The Captain was there and on the floor was a big drawn copy of the radar plot of the incident. It had to be 6 foot x 5 foot. The Captain was very amicable. The operations room was dark with red lights. He asked for my report first while my eyes adjusted, then we would look at the radar plot. I told it as it happened.

My first order was keep him abaft the beam and make him the overtaking vessel. I was laid, stopped, the only head movement I had was with keeping him abaft the beam. When collision was inevitable, I put the engines full ahead to protect my ship. Times as per my deck log book.

We all studied the plot. It showed it as I said, only I had made a little more head way than I thought.

The Captain said he was quite satisfied. We had coffee, returned to the *Lloydsman* and I never heard another thing about it.

While I was talking the whole incident over with my colleagues over coffee in the mess, a heavy smell of gas-oil enveloped the tug. We all went to investigate. On deck, forward of the accommodation, there was a fuel tank breathing pipe discharging fuel at full bore. The chief ran below and had the fuel stopped. Apparently the 3rd Engineer was trying to put 50 tons of fuel from aft into a 15 ton tank forward.

Ironically, the Icelandic coastguard plane flew over shortly after and that evening B.B.C. world news had the

'Othello'.

'Odinn'.

Lloydsman laid over in a large oil slick, seriously damaged. There were a few quick telephone calls home that night to stop the families worrying.

The Radio Officer had been on the radio with the trawler that was taking the B.B.C. TV crew home. They had tried to get the skipper to turn back but he could not lose 12 to 14 hours, his market was more important for him and his owners.

I noticed for some weeks after that the I.C.V's kept about 3 miles away from me, or at least that's the way it appeared. Perhaps the object of the exercise had been achieved. How wrong I was. The only reason we were having a quiet time was because we were in the sector the *Odinn* was guarding. With him in the dock under repairs, the only I.C.V's we saw were passing through.

CHAPTER 6

Fun and Games with the Navy

During this lull, the Navy involved the Big "L" in their manoeuvres, practising replenishment at sea being a favourite! Running a parallel course with the Royal Fleet Auxiliary Tankers or the Frigates, about 75 feet off, passing parcels, films or men by breech buoys. Usually this was done at a speed of 8 to 12 knots. One main engine on the *Lloydsman* would be ample. If more speed was required then two engines were used.

As I recall, I had no problems with the R.F.A. but on two occasions doing this manoeuvre with frigates, I would just get in position for the rocket to be fired, when my steering gear ceased operating. On both occasions it stopped with starboard helm on, away from the frigate, and off we sheared. I found out later this only happened when the frigate had his big radar on with the big "bedstead" scanner. My electric steering was being induced by his radar pulse. It was no problem, we had two power packs and it only needed the turn of a switch. But I insisted after that, he turned the "bedstead" off during all future games.

Another favourite game was to send me off with 4 hours start, to hide and the Navy would come to find me! That's a 60 plus miles start. I tried everything I could think if. I laid close up to trawlers in the middle of a pack and they found me. I would steam full speed over the horizon, make 90 degree turns, run with trawlers homeward bound and they would find me. I even crossed the 12 mile line, that was forbidden, and went 5 miles up a fjord and they found me. The only time they lost me was when I went to the 12 mile line and kept steaming down the line to the SE grounds from the NE grounds. I asked them how they did it but I always got the same answer; if I knew it would spoil the exercise. They did tell me some days later, after I had been reprimanded for poaching.

We were guarding about 15 trawlers at the Fairy River ground close to the NW corner. The trawlers' tow was in a North South track, up and down to the 12 mile line. They asked me to lay on the 12 mile line and act as a buoy. They would tow round me and out again. I agreed, it was a good place to keep a radar watch.

I mentioned before, that my radio officer was an ex-trawler R/O. His name was George Sheriff, and he had a brother who was skipper of the Grimsby trawler *Bengaum*. We were yarning on the bridge when he asked to turn the echo sounder on. It was an ordinary Simrad. We were in 50 fathoms and he was sure we were getting fish marks. If the water had not been so deep the marks would show better. I decided to test his theory. I ordered a strict radar watch and crossed the line, slow speed towards the land, looking for the 20 fathom line. A couple of hours later "Sparks" was convinced we were getting good fish marks and asked if he could tell the "boys". I told him there was no need, look aft. There they were, still towing up to me and we were only 3 miles off the land. I laid for about 1 hour. The fishing was good. My First Mate reported an echo 24 miles NE of us coming very fast, about 25 knots, I guessed it was one of our frigates and told the trawlers I had better get back to the 12 mile line.

The UHF radio was calling me long before I got there. I was summond on board. I got my reprimand for crossing the line and when asked for my comments, I said: "The said: "The fishing was good, do you want a fry?" The answer was yes, and the subject was dropped. A call to the nearest trawler was all that was needed. The frigate's boat was sent over and the trawler was only too pleased to fill it.

I asked how did he know I was the nearest to land. He showed me a plot, only one ship on it, me, no trawlers. Then he told me how he was on station 60 miles NE of the fleet, another frigate 60 miles NW of us. Between them they had cross bearings of my radar pulse. That explained how they could find me when sent to hide, and explained why they lost me once. That was the day we shut down the radar for maintenance.

I had to wait till the next Cod War before I was sent to hide again. That time I waited till I was in the pack, off radar and went to the next grounds. It took them 32 hours to find me.

H.M.S. ***Cleopatra***

When the H.M.S. *Cleopatra* was "Captain in Command Iceland", the Captain was always calling meetings on board his ship, to discuss tactics. He always sent a helicopter for me as we were usually on parameter guard duty. Normally the chopper would land on my winch house, but this particular time the weather was a bit scruffy.

All my crew were on the after end of the boat deck to watch "Dad" hoisted up on a winch wire.

Down came the winch man to give me a crash helmet, put the sling round my shoulders, tighten the toggle, and instruct me to hold the family jewels, not the sling. His last comment was watch out for the head-ache ball on the end of the winch wire.

When the wire came tight, I was leaning forward looking for it, instead of holding my head back. The ball hit me under the left eye, much to the amusement of my crew, the shower! When the winch was two-blocks, there was a man to spin you round, slack about 1 foot of winch wire, pull on the seat of your pants, allowing you to sit on the floor of the cabin with your feet dangling out board.

Only yours truly did not get far enough in board and fell out again. The slack winch wire came tight and the headache ball hit me under the right eye. Raising a second cheer from the shower.

Yours truly was the centre of amusement, throughout that meeting and the following excellent lunch. They had the right medicine to cure my headache (Bells) and by the time they returned me to my ship I was feeling no pain at all, but sporting two of the blackest eyes you had ever seen.

One day at the end of June, we were with the fleet, all fishing about 15 miles north of the Fairy River, a beautiful day, glass calm sea and not a cloud in the sky. The UHF radio burst in to life. The Navy had spotted an echo on the radar, bearing north east, coming towards us at 20 knots and ordered me to the north east corner of the fleet, which of course I did. I remember thinking, this is something new. It must be the I.C.V. *Aegir*, which was the only gun-boat that could do close to 20 knots. But they had always come at us before from the land, never the open sea.

Long before I was in my ordered position, I had him on my radar at 24 miles. The echo was large, too large for an I.C.V., it was a Norwegian cruise ship, come to watch the midnight sun. It arrived and slowly passed the fleet, gave the boys a wave, all the passengers were on deck, went about 5 miles west of the fleet, stopped, and laid there all night, lit up like a christmas tree.

I watched the sun that night myself. About 23.00 hours the sun almost touched the horizon in the NNW, stayed sat on the horizon through north to NNE and at 01.45 hours started to rise again, a new day.

A skipper called me on the VHF one day with a problem, asking if I could help. He had fouled his net and parted one of his wire bridles and his spare was a different length. The bridles are connected to the warps and the otter board, keeping the net open (I think). His mate on another trawler had a spare pair but the problem was getting them onboard. The transfer, by the mother ship's rubber boats was too dangerous because of damage to the boats by wire sprags. I told him my lifeboat could cope with that job, took the position of both trawlers and said I was on my way.

'Odinn' and Ross 'Trafalgar'

'Odinn' and Ross 'Trafalgar'

I.C.V. 'Odinn', 'Nimrod' 'Lloydsman' and Navy helicopter from H.M.S. 'Galatea'.

I closed up to H.M.S. Cleopatra and asked permission to proceed to do the transfer. Permission was refused. Captain Weir wanted me on manoeuvres. He would take care of the transfer, by helicopter. I was sent to witness and report on two frigates practising making a sea towage connection and towing each other.

I must say I was surprised at the speed they both made their connections but envious of the fact that they could use up to seventy men to do the man handling of the gear, plus two boats in the water to run the messengers. But they used what they had well and the manoeuvre went smoothly and professionally.

I learned later the helicopter transfer went okay, but the pilot had been a little worried. Apparently the helicopter winch had an automatic release at two tons and the bridles were very close to that, but all's well that ends well.

About the middle of August, the *Lloydsman* was due replenishment and crew relief. We always tried to work it so we were at the 'Whaleback', the SE corner of Iceland, so when the order came, 'Proceed to Greenock' for replenishment we were at the closest point of departure, a running start. I got permission from H.M.S. *Cleopatra* to proceed easy, east along the north coast and south along the east coast. When I got to the NE corner Langaness, there was about 30 trawlers, fishing in a 20 mile area. Two Icelandic Coastguard vessels were just laid in the middle of them. I laid between them. Apparently they had been there a couple of days. They had issued the usual warnings on the VHF "stop fishing, you are in Icelandic territorial waters", but as they did not appear to be doing anything about it, the trawlers kept fishing. I noticed how the length of time on the tows had shortened and how quicky they hauled in their gear when an I.C.V came within a mile of them, but they were working, that was the main thing, even though it was a very edgey and frustrating time, a real war of nerves. For three days we were on our toes, waiting for something to happen. Two frigates appeared on the horizon and I got the order "proceed for replenishment" on the UHF, a supposedly secure wave length. I was told not to say goodbye to the trawlers, just sneak away, if I could. It was 20.00, broad daylight so I steamed easy, north, then westerly and south along the 12 mile line.

At midnight I was about 30 miles south of the fleet when all hell broke loose on the VHF. When it got dark, (that only lasted two hours at the most), the I.C.V's had lowered their cutters and caught the boys napping. 5 warps were cut in 30 minutes. The frigates were too far away, but they were coming in among the fleet, most of which had hauled in their gear as a first precaution and to wait and see what happened.

I decided to return. I had time in hand and I was less than two hours away. Full speed 18 knots. After an hour, I had the fleet on the radar at 15 miles, when a frigate (who shall be nameless) called the British Fishing Fleet on the VHF and reported another Icelandic Gun Boat was apparently approaching the fleet from the SSW, 15 miles at a speed of 18 knots. It had to be very close to us. I dropped the ranges down on the radar and sure enough, there was a nice big echo close on our port side deep in the sea clutter. My cadet called that there was a blue light on the port side Cap. I went on the wing of the bridge and there she was, one of Her Majesty's finest, one of our frigates. I called him on the UHF, identified myself and asked: "Are you running alongside the I.C.V.?" The answer was "yes". I answered, "Keep watching her", and I put the portside "name" light on. There it was, *Lloydsman* three foot high and 15 foot long. I called him "Can you see that?" Silence, then a familiar voice, "whoops sorry Norman, but you are supposed to be 30 miles south." I could not believe it. I replied, "Ask your officer of the watch, which Icelandic Gun Boat has an after towing deck 80 feet long, like the *Lloydsman*. After all I am lit up like a Christmas tree." He replied; "Okay Norman point taken," and asked if we would refrain from letting the fleet know as he feared they would never live it down. I agreed for the young fellow's sake, I knew the Captain, the "subby" was in for a hard time without being embarrassed by the trawler lads. I also told him that to save any more confusion I would proceed to Greenock as ordered. Hard to starboard and off we went. I left the *Lloydsman* at Greenock after an eleven week voyage. My relief was Jack Goldin.

It was the end of August 1973.

CHAPTER 7
Party Night

September 1973

On this leave, I was invited to the "Skippers' and Mates' Guild" annual ball at Hull.

My wife and I attended as guests of the Skipper of the trawler *Hull City*, Fred Kirby and his good lady.

It was well attended, maximum capacity, good food and an abundance of liquid refreshment.

Trawlermen are a breed on their own, one or two rough diamonds, but the salt of the earth (or should I say sea). They looked impeccable in their evening suits and behaved admirably.

It amused me to see as the night moved on, how the cummerbunds were slipping up or down on the various figures and by 22.00 hours 35 percent of the bow ties disappeared.

I am sure an excellent night was had by all. Personally, I had to be driven home.

CHAPTER 8
Concert Night

23rd October 1973

I rejoined the Big "L" at Greenock with my crew. The ship was fully "Booted and spured" i.e. maximum bunkers, water and stores. These were the instructions from M.A.F.F. and the practice in the winter months. The weather at Iceland in the winter was "rude" to say the least, and the heavier we were, the more comfortable we were.

Personally I liked the *Lloydsman* at half bunkers. She was 2 knots faster at that weight and speed was everything on the job we were doing.

We arrived at the North West Cape, after an uneventful voyage, to find a fleet of 50 trawlers fishing an area of 30 miles, with two frigates, *Statesman* and the *Miranda* in attendance.

My patrol area was centre of the fleet to the north west edge. *Statesman* was centre to the south east edge of the fleet. The two frigates would patrol the fleet in a 30 mile circle. The I.C.V's had not been out for two weeks. The two governments were talking again, long may it last.

When I first went to the fishing grounds, I could not understand why the VHF radio was so crowded. The skippers were never off the radio. There was always one channel allocated for fishing and about another 10 for yarning and could they yarn. Every subject known to man would be discussed, sex, hobbies, families, holidays, politics, religion. There were also two channels for music.

The radio officer, being an ex-trawlerman, put me straight. He told me that when they were "on" the fish, the skipper would be on the bridge twenty hours per day. The only people he would see would be the radio operator and the cook. Everybody else would be on deck working a knife. They used the radio as a means to keep awake and keep their sanity. The more I listened to them, the more I thought he was right.

But the working channel, the fishing channel, that was different. Every skipper I had the pleasure of working with was a compulsive prevaricator of the truth. They all lied like flat fish on the fishing channel. For the uninitiated, I'll briefly explain. When the trawl comes to the surface, the skipper would wash the fish into the end of the net called the cod end. This cod end would be lifted onboard, emptied, re-tied, put over the side again and refilled from the main trawl and lifted onboard again. Each lift was a "bag", say three lifts to empty the trawl would be reported a "3 bag catch". Oh no, that would be a one and a bit catch. It seems they only told the truth when the catch was small. When they were on the fish, they wanted two or more trawls at it before anybody else found it.

It was amusing if several trawlers were close together, close enough to count each other's catch. The air would be blue. The standard excuse was: "Oh I was busy I must have miscounted."

As I said before, trawlermen are a unique breed. Two brothers arrived on the grounds, both skippers and both in "footballers class" from Grimsby. One wanted to send flowers to his wife on their wedding anniversary. The Icelanders would not allow the British fleet to use their radio stations during the conflict, except in an emergency. The only contact with U.K. was with the "big set", single side band or the morse key to Portishead.

First brother asked second brother, because his radio was un-serviceable would he send flowers and a short message for him for his wedding anniversary. No problem, on came second brother's R/O, took the address and the message over the VHF on the fishing channel. The outcome was, three days later, that lady in Grimsby received 14 bunches of flowers with 14 tender loving messages on her wedding anniversary. Coincidentally there were 14 Grimsby trawlers in our fleet three days before.

One evening, when things were quiet, regarding the I.C.V, and the fishing was reasonable, because nobody was complaining, the skippers decided it was time for a concert on the VHF radio. A channel was allocated and the time set for 19.00 hours. I was requested to act as the official adjudicator. I promptly "press ganged" three of the watch below, the two on watch and myself, 6 in all. I issued paper and pencil, with instructions to mark each act out of ten.

At 19.00 hours it started and went on for four hours nonstop. Every act could have been a paid artist at any working man's club in the land, I could not have been more surprised. The singers ranged from classics to pop, solo, duos, groups, musicians, guitars and accordians, comedians, novelty acts and monologues, one act after another, each act introduced by their skipper. These fellows had obviously done this before, a very pleasant and entertaining evening, I'm sure every trawler in the area had participated.

At about 23.00 hours, somebody said 6 bags and that, was that, I was the only one on that channel. They were all back on the fishing channel trying to find out who it was and where he was.

Somebody called me saying that the concert was over and would I pronounce the 1st, 2nd and 3rd.

Unfortunately, after each act, all 6 judges were so keen to announce how many points they had allocated, the trawler's name, the skipper's and contestant's names were all mixed up, even my sparks could not sort them out. Fortunately, there was no duplication in the "turns" so describing the winners would have to do. After a great deliberation or perhaps I should speak the truth and say an argument, name calling and blows threatened, the three acts were chosen. The third was a novelty act a quartet, bass, tenor, soprano and a yap, they sang like Manhatten Transfer but barked like dogs, one of the funniest things I have ever heard. This result was very well received on the VHF clapping, cheering and whistles.

Second we gave to a duo act that sounded just like Peters and Lee. This was accepted enthusiastically.

The winner we gave to a fellow that sang "Peoples Tree". He was fantastic. Singing without music is very difficult but to sing and improvise music, I think it was spoons, we thought he was marvellous.

When I announced the winner, the reception was about half the volume of the 2nd and 3rd and when he tried to get on the air to accept his award, someone would jam him with verbal abuse. This lasted about twenty minutes. I could not understand it and asked my sparks what was wrong. He was bent over laughing and said it's your turn next but would not explain. Slowly they eased off on the winner and started on me. Three skippers talking to each other about me went something like this: "I'm not giving him any more fish, no I'm not." "I think I'll ask him for the fry back I gave him this morning." The third, "I think I'll tell his wife when I get home."

That voice I knew and in I jumped. "Okay Fred what have I done?", and after five minutes banter between themselves he told me.

I am a Yorky from Hull. Half the fleet there was from Hull, and I awarded the winner from Grimsby.

The Navy preferred the trawlers to fish in packs. It was far easier to defend them. But the industry being what it was, some were catching fish, some were not, some wanted to change grounds, some did not.

Every other day, some skipper would ask permission to change grounds or to go off on his own. The answer was always the same, "If you leave the protection of the pack you are on your own". Some preferred this, on the theory that the I.C.V's would attack a pack because there was always someone who would keep fishing. The loaner would rely on his radar to spot the I.C.V's coming, in time to have his net up and out of danger by the time they arrived.

The Navy, because of a NATO regulation, had at all times to keep their ships bunkered to about 75% capacity. This meant they disappeared, very regularly, to take fuel off the constantly attending R.F.A. tanker.

It did not take the trawlers long to cotton on to this. Every time the Navy disappeared over the horizon, the fleet scattered. It took a week sometimes to round them up again.

It was on an occasion like this, a relative of mine by marriage, Brian Boyce, skipper of the *Falstaff*, called me and said he fancied a couple of hauls at the "coffin", about 4 hours steam south of where we were, and would I go with him. Things were quiet, no Navy about so I agreed with "Okay cousin, lead on".

We arrived there about 23.00 hours. Brian had "shot-away" and seemed quite happy. There were three other echoes on the radar screen, all slow moving and I took them to be other trawlers. I left orders to the watch to keep a careful lookout towards the land and went below.

At 07.00 hours when I returned to the bridge it was dense fog, the log book showed we had had this weather since 04.00 hours, the radar showed a group of four ships and the watch said "cousin" was one of them.

I called him on the VHF and got no reply. Thinking he was having a nap, I left him. We were too close to the land to use the VHF radio excessively.

At 12.00 hours the fog lifted and I closed up the four trawlers to find they were all Icelanders. Cousin was nowhere to be seen. When I did find him, he was back at the main pack having left the "coffin" at 03.00 hours before the fog. He thought I could see him go so did not use the VHF. As a consequence of this I had spent 6 hours guarding four Icelandic trawlers. What went on the VHF for the next 30 minutes I shall not repeat, but it amused the rest of the fleet.

CHAPTER 9

The Ice Patrol

The fleet was working approximately 30 miles NW of the NW cape. The navy called me with orders to investigate several echoes 20 miles NW of the fleet. Whoever they were, they were on or over the 50 mile limit line. Outside the Hindenburg Line.

Off I went at full speed. My chief liked to go full. He said it cleared the carbon out of the exhausts made by the many hours spent with the engines idling or dead slow.

As I approached them I could see they were five British trawlers fishing alonside the large area of drift ice about 10 miles square. The first comment on the VHF was: "The Big "L" has come to look after you again Bill". Bill was skipper of one of Boston Trawlers, a stern trawler. Bostons were owned by the same company as the tugs. A lot of comments used to insinuate we only looked after our own (Boston), and the rest came second. I assured everybody this was not true, but on this occasion to treat the comment with the contempt it deserved, I slid alongside the Boston boat to pass the time of day from the wing of the bridge.

Bill told me he was on his last tow before going home and he would dearly like to go in among the ice. His theory was the colder the water the more chance of cod. The *Lloydsman* needed the grass scraping off her bottom

'Lloydsman'

The 'Lloydsman' in ice.

'Lloydsman'.

Bad weather.

Bad weather.

Bad weather.

around the waterline. I told him I would push among the ice flows and if I could hold five knots for 30 minutes, I would come back to him and let him follow me.

The ice was as much as one foot thick in places and I was quite pleased the way *Lloydsman* handled, pushing the ice flows to either side with no effort, but the noise, I remember it to this day. It was like hitting an empty 45 gallon drum with a 7lb hammer every time we laid on a fresh piece of ice. Because the *Lloydsman* was built with 14 oil tanks below the water line, the noise of ice had an echoing effect. We got the first big bang as the ice hit the fore peak, echoed to the next tank and so on all the way aft, decreasing in sound, plus the sound of the ice scraping down each side of the ship, giving a noisy but very eerie effect.

I picked up the radio and told Bill we were coming back for him. He said; "Look aft," There he was about 10 yards behind me and the other four in line astern. They must have had faith in the Big "L". The mate asked what we would do if we stopped. My reply was: "There is no reason to think the ice is any thicker, remember we have a lot of unused power up our sleeve." But to be on the safe side I altered course, two points, closer to the edge of the ice flow.

When Bill hauled his nets he reported 200 baskets. The others reported good catches so we did it again with the same results.

The Navy called me with fresh orders and I had to leave then. I had enjoyed it but my crew were delighted to get away from that noise.

The new orders were to patrol the 50 mile limit line on the north coast from the centre of Iceland to the NW corner and back, one tug and one frigate on each station.

Apparently, both governments had started talking again on condition all defence ships were outside the 50 mile line. The Icelanders would not guarantee not to attack the trawlers during the talks, but nobody expected them to.

It was well into November now, the weather was deteriorating, the wind was NW and we had had two spells of black frost. Patrolling up and down the 50 mile line, no one to speak to or see, what a soul-destroying job that was.

A full gale hit us, just as we had reached the west end of our track line, about 22.00 hours. The wind and seas were from the north, we were 50 miles NW of the NW cape. I decided to dodge here all night to see what happened, instead of returning east down my patrol, which meant we would roll heavily all night.

At 03.00 hours I was called. The watch thought we were lolling too much and reported we were in black frost. Once on the bridge I reduced the engine revolution to minimum steerage way, then stood and watched. As the spray came over the bow, it froze almost before it hit the deck. With every deck light on, I watched for about an hour. The transformation was unbelievable. The water breaker on the forecastle head was a solid round mound of ice. Handrails normally 1 inch diameter were 4 to 5 inches diameter. The well deck was full and solid up to bulwarks, the bridge wings were almost full and solid, 8 inches of solid ice on the front of the bridge. The boat deck had its share and the after end of the after deck was solid to the bulwarks. Although every available man was on deck cracking ice, trying to keep the weight down it was obvious we were losing the battle. Decision time! I could not go into the land, so there was only one thing for it, south. I turned her round, weather behind us, a good half speed and she was comfortable, much easier for the deck crew to work.

Within the hour we came across my accompanying frigate (that I had not seen for 4 days). When I told him I had seven men working the ice he told me he had 60 ratings on deck and was still losing the fight. We agreed both ships were not rigged for these winter latitudes. I told him I was going south to get out of this weather. He was quick to say, "I'll go with you." Shortly after he called me to say we both had been ordered all the way south past Reykjavik and east to the Whaleback on the SE Corner.

Apparently all the defence ships had the same order,

Whaleback for orders. Before we got there we were ordered home to Hull.

The governments had finally come to an agreement, but how long would it last.

On 16th November 1973 *Lloydsman* docked in Hull with the other tugs.

It was mutually agreed by the British and Icelandic Governments that British Trawlers would be allowed to fish Icelandic waters for two years, but the catch had to be reduced by 50,000 tonnes per year.

The Icelandic Government insisted their fishing territorial limit was 50 miles.

The British Government would not accept it.

Result: Peace for two years, then another dispute.

Early 1974

The *Lloydsman* was built with a "towmaster" steering and manoeuvering system. This was chosen because she was a large tug, both in size and power, and only having one propeller with no bow thruster.

The "towmaster" system consisted of a Kort nozzle round the single propeller and five rudders, three on the after side of the nozzle and two foreward of the nozzle, one each side of the propeller shaft.

This was very successful. She was the only tug I had been in that could steer a course going astern without using engines to assist.

It was decided to remove the two foreward of the propeller because of aeration in the nozzle. I was not happy at the time but delighted with the outcome. She was half a knot faster.

CHAPTER 10
Hull City

8th April 1974

After the 1973 Cod War ended, the Lloydsman had been employed on commercial tows in the North Sea, mainly oil rigs, semi-sub and the jack-up types.

On this particular day we, the *Lloydsman,* were laid at Lyness at the entrance to Skapa Flow in the Orkneys. we were taking stores and water before resuming salvage station waiting for work.

At 1610 we departed in a hurry, after receiving a message that the trawler *Hull City* had been in collision in a position 20 miles SW of Fair Isle. 55 miles away from us, as he was homeward bound. Both vessels were closing each other.

We rendezvoused at 19.10 in dense fog, zero visibility but a calm sea. Moored on his port side with the biggest moorings we had, 6″ circ just in case we had to try and hold him up. The damage was on his starboard side, a very large hole in the fish room, the shape of a ship's bow.

Fred Kirby was the skipper, an old friend, and I really felt sorry for him. He had 1800 kit of Icelandic cod on board.

We crept into Kirkwall sound and anchored at 20.30, three cables off the pier.

9th April 1974

Just after midnight, a mud pilot came on board the *Hull City* and Fred took his ship, under his own power alongside the pier. The place chosen was ideal.The trawler was only afloat for one hour at high water, the rest of the time she was hard aground, upright and on an even keel, perfectly safe and out of danger.

My engineer and myself boarded him after daylight. Fred was feeling and looking a lot better after his worrying experience, now his ship was safe.

United Towing had organised divers, fitters, platers from Hull and a local engineering shop to effect temporary repairs to get him back to the Humber.

We the *Lloydsman*, had no commercial work on, so we stayed close to help in any way we could. The radio operators were on continuous watch, listening.

I remember watching from the quay, while two divers stood up to their chests in water, between the trawler and the quay, wet suits on of course, trying to fashion a plate over the bottom of the damage. A fitter and a plater issued orders from the quay wall, to bend the plate to the contours of the trawler's hull, or at least near enough to be able to use a cox's gun and shoot two or three drawing bolts into the plate and the ship's hull. These bolts are threaded and when nuts are attached, it is easy to draw the two plates together for welding.

When the diver did fire the bolt he obviously chose the wrong powered cartridge, too much power, the bolt went through the fashioned plate, through the trawler's hull, through 1800 kit of beautiful cod and through and out the other side of the ship!

As I remember, it took several days to finish the temporary repairs. I was ordered to escort Fred down to the Humber but that changed to towing a semi-sub oil rig. Fred was happy enough with the repairs, a big patch on the starboard side and a wooden broom handle through the hole on the port side. He took himself down to the Humber, the voyage being uneventful. Unfortunately his 1800 kit catch was condemned. I don't know why, we were eating it for weeks.

A short irrelevant yarn to the subject in hand.

The *Lloydsman* was anchored on standby in a bay north of Kirkwall Airport, in the Orkney Isles. We used this parking place, when we were in this area, out of work in between jobs.

The parking place was carefully chosen because we were easily seen from the Customs Office at the Airport.

Our yellow flag could be seen clearly when I was requesting a free pratique and usually meant a visit from H.M. Customs within a couple of days.

There was a wreck buoy, half way down the bay. The *Lloydsman* was anchored two cables from this buoy. Often we would watch the Kirkwall Sub-Aqua Club, round the point in their Zodiac, loaded down with empty one gallon detergent bottles, proceed to the wreck buoy and dive on it.

I could let not let them board us, because we were not clear of customs, but one day they 'stood off' and talked to me.

The wreck it seems was a two hatch timber ship, that hit a mine during the war, managed to creep into this bay but sank in 1942.

Apparently this wreck had been sold several times, the latest owners paying just £20. Each owner had had their various pickings, the engine room and propeller had gone, the bridge, anchor and chains had gone. All that was left for the new owners, the Kirkwall Sub-Aqua Club, was the soggy timber in the two holds, this turned out to be oregon pine and when dried and lightly planed, it was selling for £10 a plank in (1974).

They were lifting them to the surface, one at a time, with one gallon detergent bottles filled with exhaled air. They were arranged in a shed, close by, where they would dry very quickly because they were so hard and not as soggy as expected.

Their problem was working the planks through a small man hatch, because they could not get the main hatches off.

They wanted me to pull the wreck apart with the tug. I was thinking about it, we were sorting out old wires that would do the job, good gear could have been damaged so easily and new wires were so expensive. A new winch wire in those days would cost around £16,000.

Unfortunately we were ordered to sail, never to return. I'm sure they solved their little problem.

Icelandic vessel 'Thor' on the process of cutting the nets of German 'Flensburg" SK 124 at Halinn in Iceland.
Photo taken Saturday before 19th November 1973.
Photocredit: Morgunbladid.

PART 2

THE SECOND COD WAR

19.11.75 to 3.6.76

CHAPTER 11

The Second Cod war from the view point of defence by tugs.

Late 1975 The British Government tried to negotiate another fishing quota without acknowledging the 50 mile limit.

Result: Failure.

November 1975 All British trawlers were ordered outside the 50 mile limit by the Icelandic Government. Three tugs were mobilised *Lloydsman, Statesman 1* and *Euroman.* Three oil rig supply tugs were also mobilised as radar spotters: *Star Aquarius*, *Star Polaris*, and *Star Sirius*.

19th November 1975

The tug *Lloydsman* sailed from Invergordon after replenishment and a briefing by Mr. Peter Derham of M.A.F.F. My Defence Commander was Colin Chandler. My instructions were simply that I with the other tugs had to stand and observe, do nothing until the I.C.V's started cutting warps, then do all I could. We had had a year's experience but remember they had too! The official line was there would be no Navy about, but not to be surprised if they appeared very quickly, i.e. they would be over the 50 mile line waiting orders from the Government.

Mr. Archie Macphee of B.B.C. Radio News sailed with us. All my crew had been in the last Cod War and knew what to expect. After Archie met most of them, he remarked to me, they all seemed in high spirits. What was the reason? Simple Archie, on this contract with the M.A.F.F. we were all on one month on, one month off. Last time it was two months on and two months off.

22nd November

We arrived to find the fishing fleet in an approximate 60 square mile oblong, stretching from the SE Whale-back North up the east coast. Inside the 50 mile line but outside the 12 mile line. This was to be expected. I had learned during the last Cod War that there were only two fishing grounds outside the 50 mile line and they were small and rough, one on the NE corner outside the 50 mile patch, and one on the SE corner called Working-man's Bank. 4 warps had been cut since the 12th November, starting with the *St. Giles* and the *Ross Sirius*. Two more were cut that night, both on the edge of the pack.

Two I.C.V's were out that night causing havoc. So much for standing and observing, it was straight to work. The three radar spotters were well spread and doing a good job reporting the I.C.V's because the majority of the fleet kept fishing. The three tugs were very busy chasing the I.C.V's, stopping them from getting set for a cutting run. We were succeeding too until all three tugs were caught chasing the same I.C.V., allowing the other one to get away on his own towards a lone trawler. A radar spotter was close and tried to defend him. Unfortunately he was conned out of position, resulting in a warp being cut.

All three tug masters acknowledged we were at fault that night.

As daylight arrived so did a third I.C.V. All three steamed in the centre of the main group of the fleet, stopped and laid, 5 miles apart in a North-South line, a tug laid with each one, and the spotters were spread one north, one south and one west.

The result of this I.C.V's ploy was that two thirds of the fleet had stopped fishing. The skippers were very upset and the VHF was hot that day. Their dilemma was if they stayed here, so did the I.C.V's and some would get to fish. If they moved off the fish to look elsewhere the I.C.V's would follow and maybe nobody would fish.

The Navy arrived on the scene, very quickly, and I.C.V's gave us the run around for another 24 hours. It was as if they were showing contempt for us and deliberately

The 'Star Aquarius'.

provoking the frigates, running them close, daring them to collide. The outcome was, that when they crossed their 12 mile line and returned to port, no warps had been cut and no collisions occurred. The boys had kept fishing throughout although a few were complaining because they had had to haul too early when the I.C.V's had come too close.

Everything was nice and quiet for the next few days. The weather was deteriorating slowly from the North.

In the last Cod War, the I.C.V's usually left us alone if the weather was bad, not this time.

On this particular day, the I.C.V. *Thor* arrived. The wind was N'ly 6 with the sea to match, good visibility. For some reason he seemed quite content to steam at full speed, up and down the fleet with his cutter up. Everybody knew it only took minutes to lower it. Some trawlers hauled when he came near, some gambled.

When he got to the Southern end of the pack, my station, I fell in behind him, deciding to chase him. I knew he was slightly faster than the *Lloydsman* because I was fully bunkered and stored, but perhaps I could make better headway through head winds and sea. I was about 200 yards behind him, making the same speed, just under 18 knots, plenty of spray over the bow but no heavy water. We were working a running radar plot on the I.C.V. *Thor* and we noticed an echo coming fast upon us from the stern. It was H.M.S. *Brighton* with Commander Kettlewell. When he slid up alongside of me he called me on the UHF, and asked for a sit-rep, situation report. I explained what he had done over the last two hours and what my intentions were, mainly to keep him moving at this speed so that if he did lower his cutter it would be on the top of the water out of harms way. He agreed, then said he was going to try and slow him down, just for practice.

What happened next was a classic example of close-quarter ship handling without making contact. It was a privilege to witness it.

H.M.S. *Brighton* left my side doing about 30 knots, up the side of the I.C.V. *Thor*, across his bow, staightened up on his bow and zig-zagged rapidly throwing his stern port to starboard, starboard to port across the stem of the I.C.V. *Thor*. The stern quarters are the strongest part of a frigate's hull and in the last Cod War the *Odinn* incident proved the I.C.V. bows were suspect.

It was a fantastic sight with both ships covered in spray. The object had been achieved. He had slowed down a lot, his concentration absorbed on H.M.S. *Brighton*. I was overhauling him fast. Fire monitors at the ready.

The I.C.V. *Thor* saw me when the *Lloydsman*'s stem was within 30 feet of his stem. Someone came on his port bridge wing and shot back in, the I.C.V. *Thor* went hard to port, and headed west at full speed. Running cross the swell, he slowly pulled away from me. H.M.S. *Brighton* and *Lloydsman* escorted him to the 12 mile line before returning to station.

Commander Kettlewell and I discussed the success of the manoeuvre on the UHF but mutually agreed that perhaps we had "shown our hand". Would he fall for it again?

CHAPTER 12

The Battle of Seydisfjord

11th December 1975

All three tugs were in the vicinity, *Lloydsman, Statesman, Euroman*, plus the three oil rig supply tugs that were being utilised as radar spotters. The Navy was close but we never knew quite where.

Lloydsman was patrolling between the land and the trawlers, about 25 miles off the coast. The weather was moderate to rough westerly 6 to 8. We had not seen an Icelandic gun boat for over a week, long may it last.

According to "sods law" something had to happen and it did. My engineer reported an unfortunate mishap in the engine room, the consequence being, we had lost 90% of our fresh water. All we had onboard was 4 days supply and the evaporator was in need of repair and spares. The forecast was deteriorating, ruling out any water transfer from the R.F.A. tanker.

I called the *Star Aquarius*, one of our radar spotting tugs. John Simpson the master told me he had ballasted

The 'Lloydsman', with fire hoses at the full.

Herman Sigurdsson, member of crew, beside gun on board of Icelandic vessel 'Thor'. He shot the first shot in the 3rd Cod War at Seydisfjördur in Iceland at the vessel 'Lloydsman'. Photo taken on December 12th 1975.
Photocredit: Morgunbladid.

Gun on board of 'Thor' cleaned in Berufjördur in Iceland. Gunnar Sigurjónsson, left, and Herman Sigurdsson right. Photo taken 23rd January 1976.
Photocredit: Morgunbladid.

The damage to 'Thor' after collision with 'Lloydsman'. Photo taken in the Cod War on the 13th December 1975.
Photocredit: Morgunbladid.

down with fresh water for the winter weather, before he left U.K. and I could have as much as I needed. The only problem was how to get it. We agreed we had to find a "lee" under the coast and off we went. I told the other tugs our intentions but not the Navy.

We passed the 12 mile limit that was forbidden to the British Fleet for fear of arrest. Both tugs had set continuous radar watches. I led the way into Seydisfjord, still no "lee". My chart unfortunately was small scale, but apparently the fjord went west for 4 miles then branched into two fjords, one going NW that had no marked villages and one going SW to a town.

We decided on the "blind" fjord. As we entered this fjord the wind had increased because of the channelling caused by the high mountains each side of the fjord. I suggested, the only way we could conclude this transfer would be for the *Star Aquarius* to tow the *Lloydsman*, using my ropes and using his hoses up to my bow as that was where my tanks were. Working alongside each other would certainly cause some damage. John agreed and it worked perfectly. We were so comfortable, being towed gently up the fjord at about ½ knot against the wind that I could stop one engine. *Lloydsman* had two engines on one shaft with one propeller. I could take 100 tons of water, but agreed on 50 tons.

John and I are keen golfers. I admit golfers are like fishermen when it comes to yarning. There is no stopping them, such was the case that day on the VHF. I'm sure the Icelandic Coastguard heard every word during the two hours it took to give me that 50 tons, and we were so comfortable, enjoying to yarn, I asked for the other 50 tons.

One hour later, the Icelandic Coastguard plane circled us. That's when the cat was out of the bag. They had found us, and it was time to run. But we did not, we had one more hour's pumping left to complete the transfer. We had no information regarding any I.C.V's in the area. We were enjoying the fact that the tugs were not rolling about, the first time for a week, but because we were 9 miles inside Iceland, prudence prevailed. If we were going to be challenged, we had better be a bit nearer the open sea. So we reversed the course towards the entrance of the blind fjord and carried on pumping.

It took about 30 minutes with the wind behind us to get to the entrance of the blind fjord. John and I were still yarning, I was looking forward towards him and the entrance. He was looking aft towards me. My eagle-eyed mate spotted the masts behind the headland first. It was the I.C.V. *Thor*. He rounded the corner at full speed, flying the international code flags to stop the ship immediately. John dropped my ropes and we dropped his hoses. I ordered the second engine back on line "quick".

What happened next I thought was "odd" at the time, but with hindsight I am sure the Captain of the *Thor* thought the *Lloydsman* was broken down. Perhaps the plane had reported us being towed. It would look that way from the air.

The I.C.V. *Thor* completely ignored the *Lloydsman*. He called the *Star Aquarius* on VHF and ordered him to stop immediately. John did not answer him. He called me: "What now Norman?" I replied "Don't stop, put her in auto-pilot, full speed and get your head out of sight", and he did.

Again the I.C.V. *Thor* called him. "Stop your ship, I want to board you. You are in Icelandic Territorial waters".

Again, no answer from John and no call to me.

The *Star Aquarius* had a running start, although only 14 knots.

I had a perfect view. Both ships were only about 100 metres ahead of me and I am convinced the Captin of the I.C.V. *Thor* never saw me. He was too busy looking forward trying to get up to and alongside the starboard side of the *Star Aquarius*, to get his boarding party aboard. They were all lined up on his deck in uniforms, belts gaiters and side arms.

The *Star Aquarius* was an oil rig tug tender, very strongly built. The I.C.V. *Thor* although ice class, was lightly built, Navy style for speed.

I watched the I.C.V. *Thor* clatter alongside the *Star Aquarius*. He hit her so hard all his boarding party were knocked to their knees and the I.V.C. *Thor*

The 'Lloydsman'.

bounced off the *Star Aquarius* some 25 feet. *Star Aquarius* never moved off course.

The I.C.V. *Thor* veered off and turned to port for another try when he saw the *Lloydsman*. I was very close to getting between the two ships by then. He gave more port helm as if he had decided to board the *Lloydsman*. I knew he could not get across my bow. He must be coming for me.

I ordered men to the fire monitors. I had long dreamed of swilling his wheelhouse out, all those electronics to damage or filling up his engine exhausts.

Then there was one hell of a bang and I knew why he was doing that particular manoeuvre. He had fired his forward gun at me and he had to get his own bridge out of the way.

At this point, unbeknown to me, my Defence Commander, who had been somewhat agitated all afternoon, was in the radio room, off the bridge calling the Navy.

The Captain of the I.C.V. *Thor*, stood on the port wing of his bridge, his uniform "jacket off". I still remember that white shirt. He seemed more interested in a discussion with the crew of his forward gun than the *Lloydsman* because we were getting very close.

Then bang, the second shot, I didn't see this, but my crew told me, apparently the second shot went off prematurely because it covered the I.C.V. *Thor*'s Captain in cordite soot. According to my lads he was a perfect "Al Johnson" for the next trawler concert.

When *Lloydsman* and the I.C.V. *Thor* collided I had taken the con myself. We were on a converging course of about 20 degrees, my main concern then was to stop any attempt to board me. The *Lloydsman* had a very high forecastle head, her boat deck was higher than the I.C.V. *Thor*'s. The only place they could get onboard was the after towing deck.

I gave her starboard helm that brought my stern off him and pushed the flare of my bow on to his boat deck and bridge wing. Water was on the monitors, all ready waiting for my order. I had to wait because once they started I would be unable to see anything. I shouted to my Chief Mate, who was on the wing of my bridge: "What is he doing, where is the boarding party?" His reply amazed me. "Nobody on deck and he is pushing forward." I put the helm midships and let him, all the time he was under the flare of my bow, he was damaging himself. He must have forgotten about my bulbous bow, my lads were sure he went across it.

When he was clear, round my bow, I followed him a little, to place myself between him and *Star Aquarius*.

We had a good look at the I.C.V. *Thor* as he cleared. He showed extensive superficial damage to his bridge wing and boat deck and perhaps underwater damage. *Lloydsman's* damage report was several scrapes, minor indents and one 3 inch hole 25 feet above the water line.

The I.C.V. *Thor* ran with us until we cleared the land, he fired three more shots at *Lloydsman* from his after gun. I remember telling my crew not to worry, they were only blanks. Yet when we got back to Greenock the next time the bosun showed me a 3 inch round hole, right through the top of the port funnel, that could only have been done by one thing!

The I.C.V. *Thor* appeared to be unable to keep up with us, which made me think he was damaged underwater. He turned away and returned to his base. His only comment to me on the radio through the whole encounter was "I'll see you again". I replied "You know where I am, I'm not going anywhere".

Archie Macphee was the B.B.C. Radio reporter on board that trip. That was one incident that was reported as it was, no frills.

As both tugs approached the fishing fleet, I received a call to contact the Navy. Half expecting a reprimand for going into the fjord, I returned the call. It was my favourite frigate H.M.S. *Brighton*. There was always jovial reparteé between the Commander and myself. He wanted a full verbal report because of the gun fire, I asked him where he was when I needed him. He said he couldn't go in there anyway and started calling me Captain Kirk. I wonder why?

The next day was spent writing a full report for the M.A.F.F.

'Miranda's' boat.

'Miranda'.

CHAPTER 13

The Telegram

Just after midnight the day after the incident, a certain frigate called to inquire if I had heard the Navy news that night. When I replied I hadn't he told me the Foreign Secretary, Mr. Callaghan, had made a statement at some meeting and was reported in the "Daily Express". When answering a question on the Cod War, he replied, quote: "Both sides in the conflict are showing valour but there is no need for anyone to show their virility" unquote. This I took to be personal and it upset me very much.

I wrote out a telegram to send to the Foreign Secretary, but was persuaded by a good friend not to send it as I had involved my company by saying the tug was Britain's finest and not a prancing stallion. I was advised that it was in order to send a wire to an M.P. provided it was personal, so I did and it was personal.

CHAPTER 14

Archie Arrested

The doctor onboard the support ship *Miranda* was called to a sick man on board one of the trawlers. He wanted him onboard for treatment and observation. A boat transfer was performed by their own fast Gemini boat, later the doctor decided the man had to go to hospital in Iceland. This is allowed when the diplomatic formalities are adhered to. Yet another side to the versatility of the support ship and their crews.

Archie Macphee, being a true news hound, heard about this and persuaded the captain of the *Miranda* into taking him with them. He transferred to the *Miranda*.

A couple of days later, we heard Archie had been arrested for illegal entry into Iceland.

About a week after he left us, he was back onboard *Lloydsman*, very disillusioned. He had been under house arrest in the care of the Harbour Master at the port where he landed waiting for the *Miranda*. They would not allow him to fly home to the U.K. He had to go out of Iceland the same way he went in.

CHAPTER 15

The Nimrod Flight

18th December 1975

When we returned to Greenoock for crew change and replenishment I was a little anxious to say the least. This was the day I had to explain why I was minus 9 miles inside Iceland up a fjord.

As I moored the tug to the Quay in Greenock, I noticed three familiar gentlemen on the Quay.

Mr. P. Derham from the Ministry of Agriculture, Food and Fisheries, the Senior Surveyor Department of Trade, Glasgow, and the Surveyor from Lloyds, Glasgow.

This surprised me because I knew when my company signed the contract for the first Cod War, the ship was taken out of Lloyds and carried on the Government insurance; also any incidents that happened in Icelandic waters, were reported to M.A.F.F. alone. Anything happening on normal sailing was reported to the D.O.T.

Mr. P. Derham and the D.O.T. Surveyor came straight to my room. The Lloyds Surveyor went straight forward to the damage. Mr. Derham told me he had employed a Lloyds Surveyor because the tug was in Lloyds when the owners had it and he required a quick survey and repairs if needed, to meet the sailing time, 24 hours.

After the usual pleasantries, the D.O.T. surveyor asked for the official log book, which I gave him at the same time I gave the collision report to Mr. Derham. All the log showed was the date, time, position, and vessels involved. He asked to see the report and was told perhaps, after it had been processed at the Ministry. He smiled, had a beer and left.

The Lloyds Surveyor reported he would issue a seaworthy certificate for sailing if a patch was welded on the small hole in the hull and 2 foot of welding in the fore peak. He went to organise it.

When we were on our own I asked him straight out, "Am I in trouble Peter?" He tapped the report and said about this there is no problem, about the telegram: "I

Tug 'Euroman' *Tug 'Lloydsman'* *Frigate* *I.C.V. 'Odinn'*

Frigate *I.C.V. 'Baldour'* *Tug 'Statesman'.*

don't know, but you caused a few ripples believe me. Anyway don't worry about that. Do you have to go on the bus with the crew to Hull today?" "Why?" I said. "I'm not telling you it's a surprise. I am sure you will enjoy it. All I'll say is Archie is staying too." "Okay but I must call home first." I checked with my wife, there was no crisis at home so the surprise was on. When I told Peter he said "Oh I forgot, you have a suit with you?" "Yes" I said. "Good, I want you to travel light, send your kit on the bus." The plot thickens.

The bus arrived with the crew and Captain, my relief, about 13.30 hours. By the time we had handed over, it was turned 15.00 hours , my crew had gone, a limousine arrived and the three of us left the tug.

That was when Peter Derham told us where we were going, Glasgow Airport, Inverness Airport, car to R.A.F. Lossiemouth, stay the night as guests of the Officers' Mess. Board a Nimrod for an Icelandic patrol, back in 14 hours, night sleeper, Inverness to York — Hull for breakfast, fantastic.

We were well received that night in the Officers' Mess, by the crews of the Nimrod, a very generous and amiable bunch of men. Very interesting and for me, educational. That discussion with four different points of view, the ministry, the R.A.F., the B.B.C., and my two cents worth.

As I remember, briefing was at 04.00 hours and we took off about 05.00 hours. The Nimrod is a fascinating aircraft, I'm sure it is still covered by the Official Secrets Act so I cannot tell you how much in detail but one thing is for sure, no-one would believe how they could get so much electronic equipment in one aircraft.

After take off we flew over Scappa Flow. The British Battleships that were sunk in Scappa during the Second World War by the German submarine, were covered over with steel nets and left, just as they were, as national memorials, they are photographed each day, on a clear calm day they are easily visible. They told me that one day they spotted some skin divers diving on one memorial. One phone call and the harbour police caught them red-handed.

From Scappa up the Norwegian coast, west to the east coast of Iceland, four hours after take off and there I was talking to trawlers on radio from the Nimrod, a very memorable experience.

The Nimrod made three passes on the east coast, one from north to south, one south to north, and the final north to south again, after which they had every ship that was sailing between the 12 mile and 50 mile lines plotted on one big screen. 68 trawlers, 3 I.C.V's, and 5 cargo boats and 4 frigates with 1 R.F.A. Tanker.

The ship recognition of the crew was excellent, far better than mine, but I was able to help them once, when one of the crew remarked that the only trawlers that confused them were the foreign ones. I told him that I had never seen a foreign trawler in Iceland.

The captain told us we had time for another half run before we went home. I asked permission to tell the tugs the I.C.V. positions, the Navy had already been told but permission was granted and I was invited back to the flight deck. When we got back to the middle of the fleet *Statesman* and one frigate were engaging the I.C.V. *Aegir*. *Euroman* and one frigate were engaging the I.C.V. *Tyr.* I passed the information about the third I.C.V. to *Statesman* knowing he would pass it on to the radar spotters.

I wished them all the best and we turned for home. The Nimrod had gone low in the turn when the Captain said, "There, look, there is a foreigner," I grinned and was able to show off a little. It was one of Bostons old ships built in Poland.

We were back at Lossiemouth in time for dinner. I was taken to Inverness, caught the sleeper to York and was in Hull for breakfast as planned.

H.M.S. 'Galatea' and I.C.V. 'Odinn'.

H.M.S. 'Galatea' and the I.C.V. 'Tyr'.

The vessel 'Leander' runs into 'Thor' on 10th January 1976.
Photocredit: Morgunbladid.

Members of 'Thor's' crew inspect damage after a collision, 9th January 1976.
Photocredit: Fridgeir Olgeirsson, Morgunbladid.

The following first appeared in the *Listener* for 25th December 1975.

View from a bridge

The cod war, at close quarters, is not just a ritual game played by the trawlers, gunboats and protection vessels, according to Archie MacPhee, reporting on board the tug *Lloydsman* for *From Our Own Correspondent* (Radio 4). He was close enough to see the reality on the faces of Icelandic crewmen.

'The day I joined the *Lloydsman*, to begin what was to me, four weeks of excitement, discomfort, and at times, real fear, I remember discussing with the captain, Norman Storey, the dangers facing us. We were about to set out on a 4,000-mile voyage to waters dangerous enough in normal times, but now with the additional hazards of scheming gunboats. Although a veteran of the last cod war, and a salvage-tug master who had rescued many a ship at sea under impossible conditions, Captain Storey said: "While you're on my ship, never mention the word 'danger'." I asked why, and he replied: "I want my wife, and the wives of the men in my crew, to sleep in peace at night." So, in my reports from the *Lloydsman*, I never mentioned that forbidden word, danger. But now we are homeward bound for Christmas, I feel I'm able to report on some of the hazards the men in the British trawler fleet off Iceland have faced in producing the cod that is still the most popular item on the fishmonger's slab.

'The main enemy up here, as it has been for the 50 years or so British trawlers have been coming to Iceland, is the weather. Continual storms, with Force-ten or 60 mile-an-hour winds and over, the fear of black ice when the spray freezes in the air and builds up on a ship's super-structure, and when tons of ice have to be hacked off by hand to prevent the ship capsizing; the sea is as steep as a block of flats and coming in all directions.

But the feature I'll always remember about the Icelandic weather is its sudden changes — calm, almost mild, Mediterranean weather in the afternoon, and, at night, a furious gale and mountainous seas.

'But, with all this terrible weather, the cod run best in winter on the fishing banks off Iceland. The banks, or the grounds, encircle Iceland up to 50 miles and the 100-fathom line offshore. The trawler tows or trawls its giant cone-shaped net along the seabed, where the cod are feeding. It seems a simple enough operation, but first the fish have to be found, and when they are — usually in large shoals — they have to be caught, the skipper on the trawler's bridge following every movement of the shoal on his electronic fish-finder. Every four hours or so, a trawler shoots and hauls its net when fishing, 24 hours, night and day, when the weather allows — and for two weeks at a time or more off Iceland. It's back-breaking, cruel work and it's dangerous.

Imagine the foredeck of a trawler, iced up and rolling, gunwales under, from side to side, as I saw the other day. As the net comes in after each haul, the fish have to be gutted there and then on the open deck before they are stowed away in the fish hold. It's heavy, slippery work, gutting fish that can be up to four feet long and weigh 70 lb or more. Wrapped in thick oilskins and wearing heavy boots, a trawlerman gutting cannot wear a lifeline or lifebelt for this work. And if he does lose his footing and falls overboard, the chances of survival in these icy waters are nil. The cold will kill him in under a minute.

'But, in addition to the normal hazards of trawler fishing off iceland, there is the gunboat wire-cutter. The Icelandic coast guard service has developed this technique to scare the British trawlermen away from what they consider are their traditional fishing grounds. It's a simple technique, and highly dangerous. The trawler tows its net with two wires up to 1,500 feet long, and they go straight down to the seabed. The gunboat trails a wire of about the same length with a grapnel cutter at the end of it, crosses close under the trawler's stern, and as the cutter passes over the net, it snaps the towing wires.

'The gunboat captains have got this technique down to a fine art, and they can cut away all the trawler's gear and net in one pass. For the trawler, it means a loss of over £3,000. The dangerous part of the trawl-cutting operation is if the wire snaps back and cuts across the trawler's fore-deck. And if the cutter wire gets caught in the net, the gunboat and the trawler are locked together, and the gunboat, being more powerful, could pull the trawler over on its beam-ends, with all the consequences of that for the crewmen on deck. The Icelanders say that because of the skill — more vainly, the daring — of their gunboat captains, the practice of trawl-wire cutting is not dangerous. They point to nearly 200 wire-cutting incidents in both cod wars, and no serious accidents.

'But I have now seen wire-cutting at close quarters, and the risks the gunboat captains take in cutting close to a trawler's stern are quite terrifying. In the dark, and in bad weather, it doesn't seem to make much difference — the gunboats still go in.

'As an eyewitness, I can say quite definitely that the Icelanders are skilful enough to bring their gunboats — and their victims, the trawlers — within a hair's breadth of disaster. It would take only a momentary error of judgement for the two ships to collide, with all that means in heavy seas and strong winds.

'I saw something of this in what is now being called the Battle of Seydesfjord. It seemed from the *Lloydsman*'s bridge that the gunboat *Thor* was trying to ram us. She certainly fired her gun. Her captain tried to get around the *Lloydsman*'s bows, which he had done several times before when attacking trawlers, but the two ships collided. For the space of a minute or so, as they were locked together, I knew real fear — not for myself or my companions on the *Lloydsman*'s bridge, who had just had a shell fired at them — but for the crew of the *Thor*. As the bigger, more powerful *Lloydsman* towered over the gunboat and ground into its side. I could see that look of horror on the Icelandic crew's faces. They were thinking, no doubt, of their ship sinking in these icy waters and their nil chances of survival — the same feelings the crews of British trawlers must have when they are attacked by an Icelandic gunboat.

'So, let no one think that this cod war is a polite and danger-free dispute between two NATO allies. It is full of danger for both sides.'

CHAPTER 16

Archie's Party

29th December 1975

B.B.C. Dinner: Mr. Archie Macphee and his boss at the B.B.C. arranged dinner at the Station Hotel, Hull, for the *Lloydsman's* crew to show their appreciation for the good coverage of newsworthy items, received over the last month.

Three local skippers were also invited namely Mr. and Mrs. Fred Kirby, Mr. and Mrs. Chez Abbot, and Mr. and Mrs Eric Thundercliff. He was the Liaison Officer between the Navy and the trawlers.

A thoroughly pleasant evening was had by all, very much appreciated, good food, plenty to drink, climaxed by yours truly being presented with a small cannon, that I still have to this day.

Most embarrassing really, the cannon was chosen because of my quote: "Don't worry lads they are only blanks", when the hole in the funnel proved at least one was not.

CHAPTER 17

The Navy Withdrew

18th January 1976

I rejoined the *Lloydsman* at Greenock and proceeded back to Iceland.

A lot had happened on the fishing ground during the month we were away. Both *Miranda* and the *Othello* had landed sick men ashore to hospital in Iceland and suffered all kinds of abuse at the hands of the locals, throwing rubbish and bottles at the boat crews.

The Navy were having a hard time too. Apparently the I.C.V's had been using our tricks, hitting the frigates with their sterns causing damage to their very thin sides. The navy started to dish out the same medicine, obviously with more success because of their superior speed. The outcome was the Icelanders were up in arms over it and threatened to break off diplomatic relations altogether.

20th January 1987

The Government ordered the Navy off the fishing grounds, but where to? We did not know.

This trip we had a full complement of tourists on board. The B.B.C. TV had Peter Stewart as reporter with a sound man and a camera man. I.T.N. had Michael Oliver as a reporter. Dag Pike was aboard as a freelance journalist.

We arrived at the fleet about 30 in number, stretched out over 30 miles at the NE corner between Langaness and the 40 mile patch. The I.C.V. *Tyr* was laid with them.

The two Governments were talking. Usually the I.C.V's stayed away during talks but not this time. The Navy had been ordered out of the area during the talks. My orders were observe and report unless warps were cut.

The I.C.V. *Tyr* had all the trawlers within 5 miles of him, laid with their gear on deck. He had called each one separately on the VHF, "You are in Icelandic territorial waters, stop fishing immediately or be arrested."

The rest of the fleet were still fishing when they could. The VHF was red hot that night. The skippers were working themselves up to fever pitch and rightly so. The I.C.V. *Tyr* would not answer any calls. The skippers could not understand why they could not fish while the talks were on, as before.

I did notice one thing of significance. All trawler names were dropped on the VHF to protect the men still fishing.

The next morning the I.C.V's *Aegir* and *Baldour* arrived on the scene. The radar spotters gave us good warning and the trawlers on the perimeter were ready for them. Each trawler was called by name and ordered to stop fishing or be arrested.

The 'Euroman'.

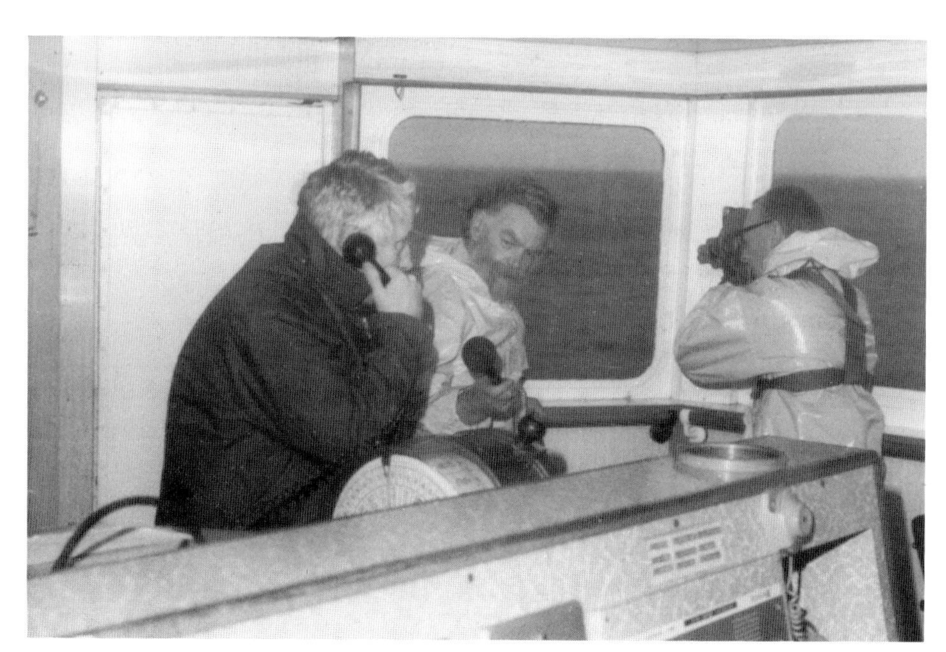

B.B.C. film crew.

B.B.C. film crew.

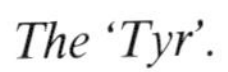

The 'Tyr'.

'Miranda's' boat.

The I.C.V. 'Aegir'.

So that was the plan. If any trawler was caught fishing after being warned, individually, he had no excuse, not only risking the costs of a warp, but the word "arrested" had crept into the threat and really upset the skippers.

The next 24 hours were pathetic. Every conceivable idea and suggestion was voiced that day. The bottom line being, how could the Government expect the trawlers to lay, indefinitely while they talked. Fishing time could not be made up and most were running out of time, being ordered back for certain markets.

On the 23rd January 1976 a vote was taken and it was carried unanimously. An ultimatum was sent to the U.K. "Send the Navy back within 24 hours or every British trawler would leave Icelandic water." Also "While obeying the orders of the British Government regarding fishing Icelandic water, compensation was required for the time they were hindered from fishing."

We spent the time slowly steaming round the fleet. I think the B.B.C. had film footage of every trawler in the area, all laid with their nets on deck. There was an embarrassing moment. I was below at the time. Peter Stewart of the B.B.C. had called the I.C.V. *Aegir* and asked him to steam past us at full speed so that the B.B.C. could get him on film. Surpisingly he answered, agreed and did.

Peter told me afterwards, they had got some excellent shots and how well he could present it, on the one hand all the trawlers laid, on the other, this vessel was part of the cause of it. He could not understand why the trawler skippers gave him two hours of abuse on the VHF, when he tried to explain why he had asked him to sail close by us.

Next day we noticed about six trawlers had gone south, either home or to the Whaleback at the SE corner. The I.C.V. *Tyr* was slowly following. The fleet had slowly elongated over 30 miles. The I.C.V. *Baldour* had gone with the trawlers that were furthest to the NW. That left the I.C.V. *Aegir* with the main fleet and the *Lloydsman.*

16.00 hours, the deadline, arrived and passed. It was dark now and I decided to stay with the I.C.V. *Aegir* for the night. Where he went, I would go, but he just laid all night, watching. I noticed that if I got inside of a half mile of him he moved outside the half mile again. Obviously, his night orders were to keep that tug at the half mile distance, all night. More discussions on the VHF.

The next day, the fleet had spread a bit more. The radar spotters had reported the I.C.V. *Tyr* had left the area going south, the I.C.V. *Baldour* was still with the further NW trawlers. It was also reported a R.F.A. Tanker had been sighted to the north, steaming west. An hour later the I.C.V. *Baldour* was reported steaming west. Question: "Was the rumour right, was the Navy just outside the fifty mile line?"

That evening, about 20.00 hours, we received word that the two Governments were still talking and we, the trawlers, had to wait, whatever that meant. The skippers did not like that at all.

The I.C.V. *Aegir* started to move. He would steam at 15 knots for 5 miles then stop for an hour. Three times he did this. Each time we stopped with him, we ran a radar plot to see if any trawlers were moving, trawling at 5 knots. I assumed the I.C.V. was doing the same, or was he just giving me the run around.

We had an echo, coming straight for me at 5 knots. 4 miles from me, 4½ miles from the I.C.V. I turned towards the trawler, nice and slow at 1 knot. The I.C.V. *Aegir* stayed where he was. He had no need to move. The trawler was towing straight towards him, the perfect position to loop him. I was reminded by my Defence Commmander of my orders: observe and report. My answer was I would observe and report, from above his warps, and I took up position above his net.

It was a Boston trawler (one of ours), a stern fisher. The skippers in these stern trawlers had voiced a lot of concern on the VHF the last few days about the possibility of a zodiac, at full speed, being able to shoot up the stern fishing ramps on this type of trawler, thus arresting the vessel. An interesting thought, but I would have argued against it.

The I.C.V. *Aegir* came on the VHF. "Stop fishing immediately" calling him by name. "This is your second

I.C.V. 'Odinn' on a cutting run, attacking 'St. Gerontius'.

The 'Odinn' and 'St. Gerontius'.

Notice the warps and the cutter. 'Odinn' going to pass.

Bull's eye. Note the cutter out.

I.C.V. 'Odinn' trying to loop the 'St. Gerontius'. Too close.
Trawler had his gear up and was able to move ahead as gun boat crossed.

warning". The skipper answered calmly with the standard answer "My Government say we are in international waters and I have a living to make and must carry on fishing. What you do is your business". That's called throwing the gauntlet down.

No reply came from the I.C.V. *Aegir*. Now we would see what his orders were during the talks. Was he there just to warn or would he cut?

I dared not say a word on the radio for fear of being accused of inciting the situation. I was about 200 yards astern of the trawler. His course was taking him across the bow of the I.C.V. We could not see if he had his cutter down on the seabed. If he had, all he had to do was move slowly ahead on to the other bow of the trawler, making him tow his trawl over the cutter. No way could I allow that.

I decided to move up to the trawler's stern. If he moved on the trawler's starboard bow I had no option but to go for him and make him tow the cutter away. If he stayed on the port bow I had to be very careful. If I went for him too soon he might get round the bow of the trawler ahead of me, then he would circle the trawler and net and do his worst.

The decision was made for me. As I closed up to the trawler the I.C.V. *Aegir* went full ahead across the bow of the trawler and was away. I was after him and I could see his cutter was not down.

While we were chasing I called the Boston Trawler and asked if he had ever had a warp cut. His reply was negative, I asked if he had heard of the looping manoeuvre. Negative again, I turned the radio to full power and told him and any other skipper listening, that the situation he had just had was perfect for looping and went into great detail on how it was performed, also recommending that as there were only 3 tugs, perhaps they should fish in pairs with one guarding.

He said he appreciated what I had suggested, but that meant only 50% fishing and everybody was well behind the clock now. Too much time had been lost already. That's why he had "stuck his neck out" as they say.

The I.C.V. ran us about for five hours, charging from trawler to trawler. They were so well scattered they had plenty of time to get the nets up. When it got dark, that's when they were most dangerous, but he decided to go home. We followed to the 12 mile line then returned to the fleet.

CHAPTER 18

"Cocker's" Bag

A couple of days later we heard the talks had broken down, but there was no mention of the Navy returning. We expected all the I.C.V's to return but we had about ten days peace. Again they were unpredictable. During the lull in the interference the fleet spread in every direction. The loners were off on their own, but most stayed in small groups of 3 or 4. The defence team was forced to spread too. The *Euroman*, *Othello* and one radar spotter had gone south to the SE Whaleback. The *Statesman*, *Miranda* and a radar spotter stayed at the NE corner. Dag Pike transferred to the *Miranda* to see for himself what the mothership operation was. The *Ranger Brisies*, now called the *Hausa*, myself and a radar spotter followed a group of six trawlers west, along the north coast.

The fishing was spasmodic until the *Real Madrid* found it. I was very close when he hauled. Daylight had just broken. "Cocker" the skipper was jubilant. He could not resist calling me on the VHF. "Can you see my bag Norm?" I could, too, and saw it shoot to the surface. I swear it jumped out of the water. There were so many fish, the net laid on the surface like a big whale. Cocker had problems getting the net alongside, but the mate had more problems getting the derrick hooked into the cod-end rope. I'm sure he jumped on the bag, walking on it to make the first connection. 13 times the cod-end was lifted onboard, a 13 bag haul, the biggest I had ever seen.

There had only been one six word comment made to me on the VHF. When he hauled the nearest trawler was

2 miles away. By the time he had the last bag onboard all six trawlers were within 2 miles of him. When he shot away the other five followed in line astern down his tow track. When Cocker hauled again, after a three hour tow, there was another good catch. There were now 16 trawlers on the patch.

They didn't miss a thing on the fishing channel, if one sneezed they all would get the 'flu.

CHAPTER 19

Tug *Euroman*

After a couple of days the fishing took off, with 5 hour tows for about 30 baskets. We had the coastguard plane over us two days running, so the skippers decided to look elsewhere. Some went back to the NE corner, some carried on west.

We kept to the east end of our little group, one eye to the shore, looking for I.C.V's, the other on the calendar. Relief day was getting nearer. My passengers were getting bored and ready for home.

There was some I.C.V. activity, but not where we were. Everytime I responded to the call, they had gone by the time I got there.

By the 7th February 1976 I had worked my way round to the Whaleback on the S.E. corner, for my last 24 hours before proceeding to Greenock for relief and replenishment.

The tug *Euroman* had just arrived from Greenock.

Built:	1967 ex. Bremen.
Length:	180′ x Beam 33′ x Depth 20′.
Gross:	1182 Tonnes.
Berths:	16 Men.
B.H.P.:	5000
Speed:	15 Knots
Bollard Pull:	75 Tonnes
Master:	Capt. Charles H. Noble.

The two Defence Commanders "handed over" on the VHF. Charlie and I yarned on our portable VHF's, only about a one mile range. We had just about finished when a voice on the big set, our radar spotter, informed everybody there was an echo. A ship had just crossed the 12 mile line coming towards the pack at 17½ knots. It had to be an I.C.V. Using the VHF it was agreed the *Euroman* would guard half the pack, centre to the west, and I would guard the other half, centre to the east.

There were twelve trawlers in the pack well spread. The I.C.V. *Aegir* kept us on the move the whole night, but seemed to favour the west side. Charles managed to give him a nudge and by daylight he was gone. 09.00 hours we were ordered to Greenock.

CHAPTER 20

London Television

11th February 1976

The *Lloydsman* berthed at Greenock for crew change and replenishment.

Peter Stewart and his B.B.C. crew seemed quite pleased with what they had on film, and they promised to send some copy footage of what went out on the television news.

I was also invited to London to see the news go out. I accepted, subject to arrangements being made.

My wife and I were invited by post. We accepted and arrived for lunch at the King's Cross Hotel restaurant. I was delighted to see all familiar faces. Peter Derham and his wife, the Defence Commander and his wife, Peter Stewart and the B.B.C. crew, even Michael Oliver from I.T.V.

We had a pleasant lunch and a memorable tour round the B.B.C., watched the work-up to the news, then the 6 p.m. news transmitted, most interesting and professional. I.T.V. invited everybody to their place the following night, but only my wife and I accepted. Another good night was had by all, but that's another story.

CHAPTER 21

The Knotted Trawlers

11th March 1976

I rejoined the *Lloydsman* at Greenock. Two things pleased me that day. The first was the tug had not been

After-deck of 'Euroman'.

The 'Euroman'.

bunkered. I was leaving with over 60% capacity and by the time we started work we would be down to 50% fuel on board. That mean't another ½ knot at least.

The second was, Peter Martin was my Defence Commander for the voyage. I knew him when he was on the *Euroman* with Charles. Ex. Navy, he knew all about R.N. procedures and would be a great help to me. He had been briefed in London before joining the *Lloydsman* and he soon brought me up to date. Apparently when the Navy returned to the fleet, the I.C.V's had been their usual defiant selves and several slight collisions had occurred. The outcome was, all British defence vessels at Iceland were now authorised to *collide* in the defence of warps. The trip back to Iceland was uneventful and we rendezvoused with *Euroman* at the Whaleback on the SE corner. He was due relief as soon as he had handed over to me.

We had about eight trawlers to nurse and to help us we had one radar spotter and the *Miranda*.

Statesman was somewhere on the north coast but had to be back at the Whaleback in eight days for his relief, when the *Euroman* returned. This meant I would be moved up to the NE corner in 4 days, then the north coast.

According to the UHF, the Navy were off on manoeuvres and nobody knew where the I.C.V's were. The fleet was fishing NE and SW over 15 miles, some 30 miles off the coast. As the spotter was on the 12 mile line, we should get ample warning, so I decided to lay in the middle of the pack and drift, anything for a quiet life.

The peace and tranquillity lasted just two days. Early on the third morning I was called to the bridge to find pandemonium on the VHF. Apparently a British side trawler had hooked his net doors and gear on an Icelandic stern trawler's doors and gear, or perhaps the other way round. *Miranda* was first on the scene, the spotter second and the big "L" last. The ministry man on the *Miranda* had the situation "sussed". When the two trawlers fouled their gear, the Brit in the side-winder "knocked out" i.e. released his bridle apparatus that holds both warps aft and together when towing. He had tried to heave and lift his gear to the surface but there was too much weight due to the Icelander still towing.

The Icelander, who could not speak one word of English, we found out later, could only see the 12 mile line, one mile ahead of him, so he kept towing, that meant the Brit was being towed sidewards from the tops of his "gallow's". Each time he laid over, the Brit gave the Icelander a lot of abuse on the VHF. Obviously he could not understand or did not want to because he kept towing. *Miranda* asked if I could stop him. I went past him, 20 feet off, at half speed, blowing a long blast on the Klaxon horn. When I was 100 yards ahead of him I stopped the horn, blew 3 short blasts and put the engines full speed astern, stopping the *Lloydsman* about 20 feet ahead of him. He stopped, and *Miranda* told me he was heaving in his warps. I just laid ahead of him while they worked at it, not very successfully either, the language problem being obvious.

About two hours later the *Miranda* called me to say the I.C.V. *Thor* would arrive in 1 hour. He would put his mate on the British side-winder to help with the language, providing the *Lloydsman* and the spotter moved at least 5 miles away. I agreed to move away but not 5 miles. 2 miles is enough and if he tried to take the British trawler over the 12 mile line or try to arrest him, we would be back. The message was passed but he never answered.

I laid for about three hours until the British side-winder sailed past me, all free and ready to go again. I followed him until he shot away. I thought the I.C.V. *Thor* would have come out to bother him, but he went the other way.

CHAPTER 22

Manoeuvres with the Navy

We received orders that night to proceed to the NE corner the following day, but not to leave this area before 09.00 hours.

At 08.30 hours that morning I knew why. Two of Her Majesty's finest appeared one each side of me, H.M.S. *Falmouth* and H.M.S. *Brighton*. I was informed, today was "playtime" and the manoeuvre of the day was to try and improve the speed of *Lloydsman* by inertia. My instructions were to steam at full speed and steer north, watch and report the speed on my walker log.They would run on each side of me and try to catch the tug in their speed "suck". I reminded them about their big radar scanners, how they sometimes affected my steering. They turned them off and I turned on both my steering engines, just in case. They wanted me settled and set before they started. About 30 minutes later I reported both engines balanced and the log speed 17.9 knots through the water.

What followed next was a pleasure to watch. Each frigate came up to me, one each side of my propeller wash. I could hear their two Commanders talking to each other on the UHF to close up to my stern. They had to manoeuvre their engines up and down in 5 revs per minute stages, When they had their bows level with my stern, the engine orders were up one rev, down one rev.

Each ship, when running free, had a speed trough on each side. On the *Lloydsman* the bow wave was just aft of the breakwater on the forecastle head. Then as the water went aft it dropped down and rose again to normal before it cleared the stern. This is called a speed-trough caused by suction, as the ship pushes through the water.

The object of the exercise was to get the bow waves of all three ships, all in line, then increase the two outside ships very slowly, hoping to suck the middle one along, without breaking the trough of water running between.

The two frigates slowly edged their way along each side of me, never more than 25 feet off the *Lloydsman*. When they were "set", both frigates bows were about 30 feet past the *Lloydsman's* bows. Obviously, the big "L" pushed more water than the frigates' streamlined bows.

They held me in the set position for about 30 minutes. My log still showed almost 18 knots, the navy insisted we were doing 18.4 knots. I remember thinking, I hoped my walker log was not broken or stuck.

The two Commanders agreed they were set and were ready for the experiment. I did ask, what the reaction of the two remaining ships would be if one broke the suction. As none had done this before we would have to wait and see.

The trials started. I could hear them raising their engines two revs at a time. When my log showed 19.5 knots they dropped the increases to one revolution at a time. It was amazing, everything was done so calm and cool, so matter of fact.

My log showed 20.4 knots, the Navy said 21.0 knots, H.M.S. *Falmouth* shot away first, H.M.S. *Brighton* 2 seconds later. *Lloydsman* almost stopped, the after towing deck almost filled with water. The increased trough, each side of me collapsed and filled when the frigates broke free, and the Big "L" sat down in the water and filled the deck. It was a good job all water tight doors were battened down.

Everyone was delighted by the experiment. The Navy wanted to try again but I declined. The trials had already cost me a clean pair of underpants.

CHAPTER 23

The Iceberg

We all dispersed. My orders were to proceed to the NE corner. Upon arrival the *Statesman* was already there with the *Othello*. The weather was perfect. I invited both Master and their Defence Commanders onboard for a handover briefing. That was my excuse. The real reason was I had to tell someone about the tests we had just done. I was bubbling.

We had a very nice couple hours.

I have had some rather strange orders, working with the Navy, none stranger than that day. My orders were to proceed to a grid chart reference, that was 30 miles NW of the Fairy River, locate, plot and monitor an Iceberg that had been reported drifting in that vicinity. 150 miles later we found it in its reported position, aground in 15 fathoms of water (90 feet). For two weeks we watched that Iceberg. It moved very little. I was expecting it to float as it melted but nothing happened while we were there.

'Odinn'.

The I.C.V. 'Baldour'.

When I arrived there were no British trawlers around but by the time I was ordered east again in approximately 2 weeks, we had a fleet of ten trawlers. We had a visit from the Icelandic Coastgard plane, every morning about 11.00 hours but no sign of any I.C.V's. There was plenty of I.C.V. activity at the NE corner and Whaleback grounds. Only one tug the *Euroman*, our slowest, was at the NE corner. Each day I would report and request to return, but always the answer was negative. Apparently if I returned it meant a frigate had to come west, and as the Navy were now authorised to collide in defence of warps as well as the tugs, they preferred to stay where the action was.

When I did return to the NE corner and the Whaleback at my relief time, we never saw an I.C.V. It made me wonder if the radios were secure, they always seemed to know what our orders were.

12th April 1976

We docked on the Tyne for relief this time. Cyril Hyam relieved me and had an eventful trip. The Big "L" got 3 or 4 mentions on the news.

12th May 1976

We rejoined the *Lloydsman* at Dundee that time. No-one realised it was to be our last trip to Iceland.

As soon as we arrived at the Whaleback grounds I was ordered onboard H.M.S. *Andromeda* to meet the new Commander in Chief, British Forces Iceland, Captain McQueen. A helicopter arrived, landed on the winch house top, just for me, no Defence Commander this trip.

The orders had definitely changed to the offensive. As soon as an I.C.V. crossed the 12 mile line, someone had to lay off a collision course and hold it. The Navy would work in pairs and box them in, then call on the open radio for any tug (or trawler) to come and ram the I.C.V. We were even issued with drawings of the I.C.V's showing where to hit them. Their most vulnerable part, was marked with a cross.

For two days the I.C.V's were never allowed across the 12 mile line. When they appeared, the radar spotters called the Navy on open radio, 2182 or VHF. By the time they crossed the line, there was a queue waiting for them. The result was devastating. I'm convinced the radios did it. The attitude of the British had completely changed. The game had ceased, it was now serious. Orders were given and accepted without jokes. It soon became apparent to the I.C.V's they were only safe on their side of the 12 mile line where the British were still forbidden to go. In fact the Navy started patrolling the 12 mile line to keep the odd enthusiast from our side crossing over.

Lloydsman's engines were never "off the boil" for 48 hours. As soon as things appeared quiet I asked permission to effect some fuel pipe repairs. The Chief wanted 3 hours. Permission was granted and I was ordered to the pack and to lay close.

As I arrived a "fry" was offered and the zodiac was down and in the water before we stopped.

CHAPTER 24

Film Show

With the "fry" came an adult film. The skipper's words were: "Something to pass the three hours away." I halved the crew as we were still steaming, telling them the films had to be over and ready to go back as soon as the repairs were finished. I barred the two galley boys from attending as they were only 16 years old. The *Lloydsman* had a large supernumerary cabin, it went across the whole ship. It could accommodate 12 passengers, 6 double cots dropped down out of cupboard recesses, 12 wardrobes and small table, when everything was stored well. The room was the real recreation space for the crew and that was where we had our cinema: A portable screen across one end, the old fashioned projector and stand fastened firm into the deck where the cot beds dropped, the old 8mm with the big reels.

The weather was not bad but the *Lloydsman* would not lay. She rolled her bins under. I took the con myself and dodged her dead slow on one engine head to the slight swell. That gave the men in the engine room a chance to effect the repairs and it minimised the risk of the projector flying across the recreation space. It had happened before.

The two boys presented themselves on the bridge at regular intervals, on various excuses, obviously for my benefit, proving they were not at the film show. I found out later one had stood in a wardrobe, waited for the lights to go out and watched the show through a 1″ gap, holding the door slightly open, the other watched the whole film backwards at normal speed while he rewound it back on the original reels. The films took two hours to run, the repairs needed three hours to fix. I wondered why the whole exercise took seven hours.

CHAPTER 25

I.C.V. *Baldour* Incident

The Navy arrived with orders as the film was going back to the trawler. I had to proceed to the NE corner with H.M.S. *Brighton* and H.M.S. *Falmouth*. I was delighted to be allocated to real enthusiasts. We did not steam as a convoy. Both frigates kept disappearing over the horizon in various directions, I just kept steaming up the 12 mile line at full speed, about 18 knots.

I spotted the echo on the radar, just on my port bow, about 10 degrees, 9 miles away, same course, same speed, and it was there all night.

In the morning, H.M.S. *Falmouth* came alongside me and told me the ship ahead of me was the I.C.V. *Baldour*. H.M.S. *Brighton* was away bunkering and would be here with us about noon. Then they intended carrying me a bit closer. He had a theory that perhaps the *Baldour* was going full speed because he thought he was undetected. "I will just show him he is not," he said laughing.

H.M.S. *Falmouth* shot away in the mist at about 29 knots. 45 minutes later as I watched my radar, I saw him pass the I.C.V. close and keep going straight past him, due north and maintaining his speed. Sure enough, as the frigate left him I.C.V. *Baldour* reduced speed, down to 15 knots.

In two hours I was within 3 miles of him getting closer all the time. Then my luck ran out. The visibility improved a lot and he could see me. Immediately, he went back to full speed but held his course. There we were almost 3 miles apart and I was slightly faster, about 0.2 of a knot. It would take 15 hours to catch him, most frustrating.

About 14.00 hours H.M.S. *Falmouth* and H.M.S. *Brighton* came up on me from the stern. I remember that surprised me. I had been so concerned with the job in hand and what I was doing, both radars on low ranges, one on 6 miles and one on 12 mile ranges oblivious of what was happening on the 24 mile range and above.

H.M.S. *Falmouth* had taken a long slow 30 mile circle turn, rendezvoused with H.M.S *Brighton* and came up behind me.

Both the frigates were soon in position, one each side of me, much quicker than when we practiced. In two hours we were within half a mile of the *Baldour*, then I assume he thought discretion was the better part of valour. He had held course 2 miles outside the 12 mile line, steaming north all the time we closed him, then he went hard to port 90° to his course, over the 12 mile line, where we all could see an extensive fog bank.

We lost our inertia, trying to turn too quickly and although the frigates picked me up again we were by then over the 12 mile line in a thick pea souper. Visibility was far too bad to play the game we were taking part in. My two companions turned back to the ordered course, and I chased for a few minutes more. We were so close but I could not see him, then the UHF radio calmly said, "Come out of there Norman, your course is north." What a nice way to get my orders, so out I came. We went north but the I.C.V. *Baldour* went south.

When I found the trawler fleet they were about 20 miles off the NE Corner. No sign of the Navy. They were on the 12 mile line.

Several uneventful days passed, although plenty was

going on at the 12 mile line. The overseas news was full of it, incidents happening most days.

We were buzzed by a Navy helicopter one forenoon. The weather was good and he requested landing on the winch house, granted of course. It turned out to be the old man himself, the Commander in Chief, Captain McQueen. "Just thought I would come and let you know what's happening," he said.

Obviously very pleased with the way things were going, he said no I.C.V. had crossed the 12 mile line for 3 weeks although they had tried most days.

He was about to leave when he said to me, "I hope you don't mind me saying but you have a wire dangling down over your bow." "Not at all Sir," I replied with a big smile, "that's my secret weapon. There is a cutter catcher on the end of it. When the I.C.V's have a cutter out and they think I am trying to hook it, they must keep moving. If they are motoring they cannot cut warps."

Then he told me about their secret weapon. They have a machine that could send out a pulse, lock on to any radar it was aimed at and render it useless. I could not resist saying, "Why don't you park outside each Icelandic port and when an I.C.V. comes out, a quick blast and he will go straight back in for repairs."

Apparently the arc of the beam is not narrow enough and every radio and television behind the target would also suffer. Radios in Iceland are used for medicine and doctors in the remote outback. That's why the Navy never used it.

The Commander in Chief returned to his ship and that very night, according to the B.B.C. overseas news, the Icelandic Government wanted to talk. The Norwegian Government had condemned the whole affair, saying it would escalate and lives were in danger. They said peace at any price.

30th May 1976

We were ordered to a position off the Faroes Isles to wait for orders. Then to Dundee for orders.

3rd June 1976

All support tugs and supply boats docked at Dundee waiting orders.

We laid for 5 days before we all returned to normal duties. We knew it was over but I thought, just until the next time.

Some time later the Icelandic Government proclaimed their territorial fishing limit was now 200 miles.

Norway and half the world followed to 200 miles. United Kingdom stayed at 12 miles. Although many years later, we did extend to 200 miles but the European agreement allows members to fish up to each others' 12 mile limit.

In 1976 over 150 side trawlers sailed out of Hull, fishing Iceland and Norway. Today they have all gone. The fishing industry in Hull has been decimated along with all the offshoots of this great industry.

If one argued against gun boat diplomacy what happened to the diplomats? Today more Icelandic Trawlers land in Hull than anyone else, maybe 80%.

As Alice said, quote "Curious and Curiouser".

Shortly after, all the trawlers left Iceland. The Icelandic Government honoured all their gun boat Captains by awarding them "Knights of the Falcon", an order similar to our "Order of the British Empire".

In 1977 our Goverment graciously awarded two honours. Mr. Peter Dereham of the Ministry of Agriculture, Food and Fisheries with the O.B.E. and Captain Charles H. Noble, Master of the tug *Statesman* with the M.B.E.

I had to wait until June 1987 to be awarded my O.B.E. for services to the offshore oil business, but that is another "story" mentioned in my autobiography that hopefully follows this publication.

Crew relief day at Greenock, August 1973
Mrs. Maisie Storey, Mrs. Corol Whiteley, Bill Bridges, Defence Commander, with part of the "A" team. Danny Betts, Mate, George Sherrif, 1st R/O, John Davis, Bosun, 2nd R/O, A/B and greaser.